INTRODUCTION TO
the Simulation of Dynamics Using Simulink®

Chapman & Hall/CRC
Computational Science Series

SERIES EDITOR

Horst Simon

Associate Laboratory Director, Computing Sciences
Lawrence Berkeley National Laboratory
Berkeley, California, U.S.A.

AIMS AND SCOPE

This series aims to capture new developments and applications in the field of computational science through the publication of a broad range of textbooks, reference works, and handbooks. Books in this series will provide introductory as well as advanced material on mathematical, statistical, and computational methods and techniques, and will present researchers with the latest theories and experimentation. The scope of the series includes, but is not limited to, titles in the areas of scientific computing, parallel and distributed computing, high performance computing, grid computing, cluster computing, heterogeneous computing, quantum computing, and their applications in scientific disciplines such as astrophysics, aeronautics, biology, chemistry, climate modeling, combustion, cosmology, earthquake prediction, imaging, materials, neuroscience, oil exploration, and weather forecasting.

PUBLISHED TITLES

PETASCALE COMPUTING: Algorithms and Applications
Edited by David A. Bader

PROCESS ALGEBRA FOR PARALLEL AND DISTRIBUTED PROCESSING
Edited by Michael Alexander and William Gardner

GRID COMPUTING: TECHNIQUES AND APPLICATIONS
Barry Wilkinson

INTRODUCTION TO CONCURRENCY IN PROGRAMMING LANGUAGES
Matthew J. Sottile, Timothy G. Mattson, and Craig E Rasmussen

INTRODUCTION TO SCHEDULING
Yves Robert and Frédéric Vivien

SCIENTIFIC DATA MANAGEMENT: CHALLENGES, TECHNOLOGY, AND DEPLOYMENT
Edited by Arie Shoshani and Doron Rotem

INTRODUCTION TO THE SIMULATION OF DYNAMICS USING SIMULINK®
Michael A. Gray

INTRODUCTION TO
the Simulation of Dynamics Using Simulink®

Michael A. Gray

CRC Press
Taylor & Francis Group
Boca Raton London New York

CRC Press is an imprint of the
Taylor & Francis Group, an **informa** business

A CHAPMAN & HALL BOOK

MATLAB®, Simulink®, and Stateflow® are trademarks of The MathWorks, Inc. and are used with permission. The MathWorks does not warrant the accuracy of the text of exercises in this book. This book's use or discussion of MATLAB®, Simulink®, and Stateflow® software or related products does not constitute endorsement or sponsorship by The MathWorks of a particular pedagogical approach or particular use of the MATLAB®, Simulink®, and Stateflow® software.

CRC Press
Taylor & Francis Group
6000 Broken Sound Parkway NW, Suite 300
Boca Raton, FL 33487-2742

© 2011 by Taylor & Francis Group, LLC
CRC Press is an imprint of Taylor & Francis Group, an Informa business

No claim to original U.S. Government works

Version Date: 20110720

ISBN-13: 978-1-4398-1897-8 (hbk)

Library of Congress Cataloging-in-Publication Data

Gray, Michael A.
  Introduction to the simulation of dynamics using Simulink / author, Michael A. Gray.
    p. cm. -- (Computational science series)
  "A CRC title."
  Includes bibliographical references and index.
  ISBN 978-1-4398-1897-8 (hardcover : alk. paper)
  1. Dynamics--Computer simuLation. 2. SIMULINK. I. Title. II. Series.

TA352.G73 2010
620.1'040113--dc22   2010021405

Visit the Taylor & Francis Web site at
http://www.taylorandfrancis.com

and the CRC Press Web site at
http://www.crcpress.com

This book is dedicated to my wife, Mary Teresa, and my father, John M. Gray, without whose influence, constant encouragement, and unfailing support it would not have been possible.

Table of Contents

Preface

The importance of simulation as an integral part of the analytical methods of science and technology grew rapidly in the latter half of the last century, and as the twenty-first century unfolds, simulation will expand its role as a tool for modern science and technology. Because this subject is so vital in technological work, it is important that undergraduate students receive an early introduction to it so that they are comfortable with employing simulation whenever it is useful, both in their course work and later in their jobs.

To support early study of simulation, we must have undergraduate-level textbooks that are satisfactory for use by a general community of science, engineering, and technology students. Some textbooks are based on constructing simulations using general-purpose programming languages, and such books are therefore restricted to undergraduates with programming experience. Other textbooks focus on simulation and modeling in the abstract, and these typically provide few modern computer tools for the students to use. Many textbooks introduce simulation only at an advanced undergraduate or graduate level in connection with a specific scientific, engineering, or technological field. This textbook follows a different strategy by aiming at the general scientific, engineering, and technological undergraduate student community with no programming experience.

To support this strategy, I have based this study of simulation on the use of modern graphical simulation tools. The tool I use is the Simulink® software, which is part of the MATLAB® software system. The graphically based Simulink application is attractive to modern undergraduates, who have grown up on visually appealing and intuitively useful computer applications. I also stress the generality of simulation by using examples from different science, engineering, and technology fields. Early undergraduate students will benefit from this approach, since early undergraduate study is exactly the time when breadth of knowledge is sought and when unifying themes across the sciences and technologies are studied.

While the translation of a physical model into a simulation may be easy for advanced students, it is not always intuitive for beginning undergraduates, so I concentrate on giving the reader a clear explanation of how to go from physical models described by mathematical equations directly to executable Simulink simulations. The book is written in an informal style, with many diagrams and graphics, and the exercises are embedded in the body of the text so that they can be done on a learn-as-you-go basis. Chapters 1 to 9 contain the basics of building a simulation in Simulink, while Chapter 10 has an introduction to some advanced topics. This text is appropriate for a one-semester course in the topic and should prepare the student for advanced and possibly specialized studies in simulation. Instructors can be from any science, engineering, or technology field, so this book can be used in the undergraduate curricula of many different departments.

The book assumes that the student version of Simulink is the sole vehicle for constructing simulations and that no general-purpose programming skills are required. The drag-and-drop graphical interface provided by Simulink makes the construction of a simulation almost like video gaming, and the student is provided with immediate feedback and satisfaction when developing a simulation.

I start the textbook with finite-difference equations rather than ordinary differential equations (ODEs), because I believe that beginning students grasp the simulation of first-order difference equations easier than they do first-order ODEs. Simple discrete models like annual population models are easy to understand and allow the textbook to introduce the concept of *state* (in this case, the previous population value) in a simpler way than is the case in differential equations. When students move to first-order ODEs, they already understand the idea of retaining and using state, which makes the task of understanding numerical integration algorithms easier. This also allows the discussion of the idea of simulation time step in a simple context, and it paves the way for understanding more complex time-step concepts later.

Finally, I would like to stress the enjoyment that simulation brings to the learning experience of the undergraduate student. Although there are many graphic demonstrations shown in courses by instructors, the satisfaction of personally constructing a simulation that performs as expected is very thrilling intellectually. The simulation then allows students to explore behavior and effects beyond the standard course topics, which deepens and enriches their understanding of the simulated systems.

CHAPTER 1

Introduction and Motivation

O NE OF HUMANITY'S GREATEST intellectual achievements happened when humans first realized that it was not necessary to use a real, physical system to analyze changes to it. People discovered that a small replica of a system could be constructed to investigate possible changes without the large expenditure of effort or resources that changes to a real system would require. These replicas were our first *models*, and they have been central to science, engineering, and technology ever since. Rubinstein (1981) quotes Rosenbluth and Wiener (1945), who wrote:

> No substantial part of the universe is so simple that it can be grasped and controlled without abstraction. Abstraction consists in replacing the part of the universe under consideration by a model of similar but simpler structure. Models ... are thus a central necessity of scientific procedure.

Models and their manipulation have become even more necessary as we tackle systems of global size or with a long time span. Systems whose dynamics unfold only once during the universe's lifetime or which are inherently self-destructive are things that cannot be reproduced in the laboratory at all. Accurate models that enable us to predict the future development of such systems are invaluable in these cases.

This textbook contains a study of one important technique for using models to investigate the dynamical behavior of physical systems. This technique is called *simulation*, and it consists of the construction of a model that operates in simulated time to show us the predicted dynamics of the system. There are many kinds of simulation in the world, including simulations of systems from arts, humanities, social sciences, and physical sciences, but we generally restrict our work to simulations of physical systems that arise in science, technology, and applied mathematics. We also focus mainly on physical systems whose behavior unfolds continuously in time rather than at discrete points in time, although we do examine some discretely unfolding systems for educational purposes. The models we use are mathematical models, consisting of sets of mathematical equations that predict parameters of importance for the systems.

There are many kinds of physical systems whose behavior is known to be *stochastic*. By this we mean that the system equations contain parameters whose values do *not* vary in time in a *deterministic* manner. The next value of such parameters at any time cannot be determined solely by the current time and the system's past history, but must be chosen by knowing the statistics of the processes and choosing values according to the appropriate statistical distribution. Stochastic systems require a different set of techniques from those that we study in this text, so we restrict our study to systems whose models are deterministic in nature.

1.1 SYSTEMS

To begin our study of the simulation of continuous systems, we must start with a definition of what we consider to be a system. *A system is a collection of associated parts whose combined behavior can be viewed as the behavior of a unified entity.* We need to examine the features of this definition carefully because each carries important implications about the assumptions in this definition.

> *A system is a collection*: This phrase implies that we accept the size of a system to be any number from 1 to an infinite number. So number alone is not a characterizing feature of a system. The size of a system is not a limitation on the kinds of things that can be systems, nor does it provide a litmus test for identifying systems.
>
> *of associated parts*: This piece of the definition implies that systems are composite: They have parts that are identifiable individuals. In fact, the parts of a system may themselves be systems, which we may call *subsystems* of the system. But since each subsystem is itself a system, this is a recursive definition allowing nesting to any level. Of course, there are very simple systems, which may have only one or two parts and so have a flat internal structure, but we do not make

the restriction that system structure must be either flat or nested. Note that the parts making up a system cannot be unrelated, so an arbitrary collection of things does not constitute a system. There must be some relation that associates the parts and provides us with a reason to consider the group as constituting a whole.

whose combined behavior: A crucial aspect of a system is that its parts interact to provide the system with its behavior. The parts possess individual behaviors, and they can be quite different from that of the system in which they participate. But their behaviors mesh together in such a way as to give an overall behavior to the system that is distinct from its parts.

can be viewed as the behavior of a unified entity: The last and possibly the most important aspect of the definition states that we can treat the system as a whole and base all our analyses on this black-box behavior. We are able to replace the complicated combination of part behaviors with a single behavior that we call the system behavior.

1.1.1 Examples of Systems

Systems are so pervasive in our world that we are overwhelmed with examples. Our farms are intricate *agricultural systems* for producing large amounts of food in a reliable and sustainable manner. Our food is often prepared in chemical facilities that are complex *chemical systems*. We live in an environment of animals and organisms that constitutes a *biological system* of immense complexity. Humans organize themselves into large *social systems* containing complex subsystems such as the mechanism through which we organize our trading, our *economic system*. *Mechanical systems* such as buildings and automobiles surround us in everyday life. The parts of a building—its rooms, stairways, heating and cooling mechanisms—are quite different from each other but are designed to work together to provide the overall function of sheltering us from weather. The heating and cooling mechanisms are usually subsystems due to their own systemic natures. *Electrical systems* include the power grid that delivers electrical power to our buildings. Modern cities, which are systems consisting of an extremely complex collection of physical, social, political, and economic subsystems, are very likely the most complex things humanity has ever devised. All of these systems are candidates for simulation so that we can understand their workings better and control these systems for our benefit.

1.1.2 Classifying Systems

We want to study systems in an orderly and systematic way. To do this, we need to divide them into categories according to some fundamental feature. Since all systems possess behavior and change as time goes by, we can group them according to the way in which this change occurs. Systems can be classified into *discrete-time systems* and *continuous systems* according to whether their change occurs discretely or continuously.

Discrete-time systems are systems that change only at certain times, which are the discrete times to which the name refers. Often the discrete times at which change is allowed occur in a regular, periodic fashion. This is usually the case with electrical systems, which are often controlled by a system clock. Electrical systems of this kind are called *synchronous systems*, since they are synchronized by the ticking of a clock. Computer processors are good examples of synchronous systems, and their performance is usually characterized by their clock speed expressed as a frequency in gigahertz. A classic biological discrete-time system is an animal species whose population change occurs only once a year according to a natural reproductive cycle. The discrete times of such a system are regular yearly times. Sometimes the discrete times of system change are irregularly spaced and connected with an event. These systems are sometimes called *discrete-event systems*. A classic example of a discrete-event system is a shopping-mall parking lot, whose slots are filled or emptied at irregular times, determined by the irregular arrival and departure behavior of shoppers.

The second kinds of systems studied in this text are continuous systems. A continuous system is a system whose behavior varies smoothly and continuously as the fundamental system variables change. Time is usually the fundamental independent variable for these systems, and our mathematical equations describe the change in these systems as time increases.

1.2 DYNAMICAL MODELS OF PHYSICAL SYSTEMS

Systems of the kind we have listed here usually have models consisting of *dynamical equations* describing how the system parameters of interest change as the independent variables change. The models we use for simulation of a system are the dynamical equations that are formed from the physical laws that describe the system. These usually involve relationships between rates of change of parameters and important functions of the independent variables. In this text, we study models formed from two kinds of dynamical equations: *difference equations* and *differential equations*.

1.2.1 Discrete-Time Models

A discrete-time model usually is a set of difference equations describing the evolution of the system behavior in time. A difference equation is an equation that relates the value of some fundamental system parameter at one time to the values of the parameter from earlier times, and can possibly include constants and other quantities

1.2.2 Continuous Models

Continuous models usually consist of a set of algebraic or differential equations describing the evolution of the system behavior in time. The dynamical equations needed to describe continuous physical systems often involve the rates of change of time-varying parameters and functions. The incorporation of such terms in the equations makes them into *differential equations*, so that the techniques of integral and differential calculus must be used to analyze the equations.

1.3 CONSTRUCTING SIMULATIONS FROM DYNAMICAL MODELS

Once we have models of systems in the form of dynamical equations, we can build simulations based on the models. The simulations we will study in this text are *digital simulations*—digital computer programs whose operation provides us with the simulation outputs. A simulation construction consists of either the development of a general-purpose-language program that simulates the system or the assembly of a simulation by using a simulation system that provides the simulator with a set of generic programs out of which the simulation may be constructed.

The first technique is often difficult and slow, although it provides the simulation developer with a very deep insight into how the simulation program itself works. The second technique is easy and fast, and it provides the developer with a simulation very rapidly at the tradeoff of less knowledge of the internals of the simulation code. As stated in the Preface, our goal in this text is to encourage students to use simulation from the very beginning of their study as an exploratory tool and a means of gaining intuition about system behavior. So, in this text, we will study the construction of simulations by using a simulation system—the Simulink® software that is part of the MATLAB® software environment, a widely used scientific and engineering computation system.

1.3.1 Block-Diagram Models

Block diagrams are a widely used representation method for building models in the engineering and scientific fields. The block-diagram approach consists of producing a graph, or network, of blocks representing the parts of the intended system. Blocks are chosen to represent the parts of the system, and the arcs between the blocks are chosen to represent the communication interconnections between the system parts. As an example, consider the block diagram in Figure 1.1 that shows a high-level block design of a simple computer. In this diagram, the block labeled *memory* stands for a generic memory element, with the number of addressable elements and the number of bits in an addressable element unspecified. The designer customizes the memory element by setting these properties for the element. In the computer-design field, some design elements, such as the memory element, are so common that they become generic elements that can simply be inserted into the design with just the customizations added as explanation.

Another example, shown in Figure 1.2, is a block diagram of the control of the arm movement of a disk drive. We simulate this diagram in a later chapter.

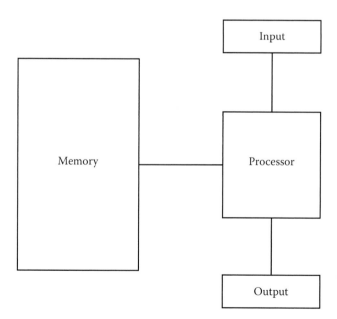

FIGURE 1.1   A block diagram of a simple computer.

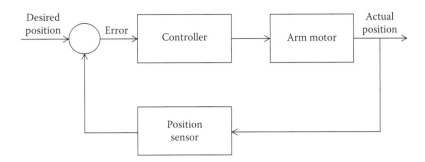

FIGURE 1.2   A block diagram of the control of a disk drive arm.

1.3.2 Block-Diagram Simulations

When a simulator works from a model consisting of mathematical equations, blocks can be used to represent terms, or sets of terms, in the mathematical equation, and the communication interconnections represent the combination of the terms. When finished, the network of elements represents the dynamical equations of the model. If simulation software provides preassembled modules for the terms, the simulation is much simpler and more intuitive to construct than when dealing directly with mathematical equations by mathematical solution techniques. In particular, focus is placed on the parameters of the elements and the design choices for these parameters and on the combinations and interconnections of the design elements.

We use the common generic elements provided by the simulation software to create a diagram of our model, and, when we have completed our customization of the elements, the simulation software will run the simulation and present us with the desired outputs. In the next chapter, we begin an exploration of how block-diagram simulations can be built in Simulink.

1.4 HOW SIMULATORS ARE USED

The classic way of using a simulation of a model is to analyze a system whose model is too difficult to analyze using symbolic methods. By symbolic methods, we mean analytic mathematical methods that provide exact or approximate solutions to the model. For simple models, such as those studied in many courses, the symbolic approach is very successful. But larger, more complicated models are often not analyzable by symbolic methods, so scientists and engineers must turn to other approaches. When a symbolic solution is not possible, it is common to try an approximate solution by reducing the actual model to a less complicated one that can be solved by symbolic methods. But, even if the approximation approach fails, digital simulation can still provide a solution. Consequently, simulation is often used as a means of last resort when analyzing a model. Of course, most of the simple scientific problems and engineering projects have been analyzed thoroughly and are well documented in the research and engineering literature, so new projects often require simulation as their analysis method.

A second way in which simulation is used is to demonstrate the features of a model's dynamics in a very explicit way to help analysts gain insight and intuition into the system's behavior. Simulations often use powerful graphics with extensive parameterization so that the effect of the parameters can be explored very rapidly. These what-if analyses can be very helpful in formulating other models that may be more useful in understanding a system's behavior.

In this book, we examine models that show both of these usages.

1.5 SUMMARY

We have discussed the importance of models in dealing with systems and their dynamics. Simulation is a main tool of the twenty-first century for analyzing models, and these models provide scientists and technologists with the opportunity to examine the dynamical behavior of models and to change physical systems.

A system is a collection of associated parts whose combined behavior can be viewed as a whole, and systems can be grouped into discrete-time and continuous systems. Discrete-time systems change at specific times, while continuous systems change smoothly and continuously.

Block diagrams are representations of systems in the form of a network of design elements represented by blocks. Simulations can be constructed from block diagrams using simulation software, and in this text we use the Simulink software of the MATLAB software environment.

REFERENCES AND ADDITIONAL READING

Bender, E. 1978. *An Introduction to Mathematical Modeling*. New York: John Wiley.

Dym, C. 2004. *Principles of Mathematical Modeling*. New York: Elsevier Academic.

Fishwick, P. 2007. The Languages of Dynamic System Modeling. In *Handbook of Dynamic System Modeling*. Ed. P. Fishwick. Boca Raton: Chapman & Hall/CRC.

Gershenfeld, N. 1999. *The Nature of Mathematical Modeling*. Cambridge: Cambridge University Press.

Huckfeldt, R., C. Kohfeld, and T. Likens. 1982. *Dynamic Modeling: An Introduction*. Beverly Hills, CA: Sage.

Kalman, D. 1997. *Elementary Mathematical Models: Order Aplenty and a Glimpse of Chaos*. Washington, DC: Mathematical Association of America.

Mooney, D., and R. Swift. 1999. *A Course in Mathematical Modeling*. Washington, DC: Mathematical Association of America.

Rosenbluth, A., and N. Wiener. 1945. The role of models in science. *Philos. Sci.* 12 (4): 316–321.

Rubinstein, R. 1981. *Simulation and the Monte Carlo Method*. New York: John Wiley.

CHAPTER 2

The Basics of Simulation in Simulink

I N THIS CHAPTER, WE discuss the basics of constructing simulations of systems in Simulink® software. Since we are just interested in learning the process of developing simulations with the software, we simulate systems that are so simple there can be no confusion about how they behave and what their simulation results should be. This allows us to focus completely on the structure of the simulations and the process for building them. We also proceed through this chapter at a slow pace so that all students are given ample time to adapt their understanding to the general process of simulation.

2.1 SIMPLEST MODEL TO SIMULATE

The simplest system to simulate is probably a system that does not change from its initial configuration—it doesn't have any significant dynamics at all. Consider a small object at some point $(3,-1,10)$ in a three-dimensional Cartesian coordinate system. If the object is at rest in the coordinate system and there are no forces acting on the object, it will remain at rest. The simulation of this system is so easy that it gives us no difficulty at all. The system is described by a model consisting of three algebraic equations, $x(t)$, $y(t)$, and $z(t)$, giving the values of the object coordinates as shown in Equations (2.1)–(2.3):

$$x(t) = 3 \qquad (2.1)$$

$$y(t) = -1 \tag{2.2}$$

$$z(t) = 10 \tag{2.3}$$

To create our simulation, we need to put these three equations into the form of a block diagram. Recall from Chapter 1 that a block diagram shows elements connected by arcs indicating value-transmission links between the elements. In our block diagrams, the elements will be *simulation blocks*. These are blocks into which inputs are supplied, processing occurs, and from which processing outputs are produced. The arcs indicate input and output connections, along which the input and output values flow. The inputs required by a block are received on incoming arcs, and the outputs produced by a block are supplied on outgoing arcs. Based on this picture, we expect that data supplied to a model from the outside appears as blocks with no input connections, while outputs of the model appear as blocks with no output connections. The flow of inputs and outputs in a block diagram is usually shown proceeding from left to right in a layout like that of a page of text, so that the outputs of the model are located on the right-hand side of the diagram.

In contrast to the layout of block diagrams, mathematical equations are laid out as *relations* between expressions. Equality relations such as those shown in Equations (2.1)–(2.3) state that the left-hand expression has the same value as the right-hand expression. In our case, the model uses these equality relations to assert that the left-hand result parameter is the same as the right-hand expression, so we think of these equations as having the output on the left-hand side. Of course, mathematical equations viewed this way flow from right to left, opposite to the usual direction in block diagrams. If we reverse the mathematical equation and change the equality symbol to an arrow, we have a picture of simulation results flowing from the expression on the left to output on the right, and the blocks that we need in our diagram become much clearer.

$$3 \rightarrow x(t) \tag{2.4}$$

$$-1 \rightarrow y(t) \tag{2.5}$$

$$10 \rightarrow z(t) \tag{2.6}$$

We need two blocks to represent each of Equations (2.4)–(2.6), with a constant as an input block and the coordinate value as an output block.

2.1.1 The Feedforward Block Diagram

Note that the inputs are generated on the left-hand side of the block diagram and pass progressively to the right-hand side with no reversals. This kind of diagram is called a *feedforward* diagram, since the values flow only in the forward direction (inputs to outputs). Feedforward diagrams are particularly easy to simulate, as we will see.

2.2 MODELS IN SIMULINK

We want to convert our block-diagram model into a simulation diagram that can be executed on a computer, so it's time to consider the simulation tool we will use throughout this text—Simulink, which is part of the MATLAB® scientific and engineering computation application. Consequently, the computer must have MATLAB software as well as Simulink software installed. In this text, we center our discussion on Simulink running on Windows XP computers, but the comments and advice generally apply to other operating systems as well.

To start Simulink, we must first start MATLAB (see Appendix B for the basics of running MATLAB). When MATLAB is running, click the Simulink Launcher icon to start Simulink. At this point, we can minimize the MATLAB window to reduce the complexity of the desktop, but we can't close it because our Simulink session will also close. Launching of Simulink causes the Library Browser window to appear, which is the window from which all work is begun.

Figure 2.1 shows the Library Browser that contains the usual menu and toolbar at the top, with an information panel just below. Below the information panel are two columnar subwindows showing a list of the Simulink libraries on the left and the blocks in the currently selected library on the right. Before we start to use the blocks, we must create a new model in which our simulation will be placed.

We select the File | New | Model menu item, and this causes Simulink to open a new Model Editor window similar to the one shown in Figure 2.2. This window is the Model Editor that allows us to insert blocks representing the blocks in the block diagram.

2.2.1 Documenting a Model

In developing a complex simulation, we can expect to create a number of versions, so it is very important to document the models properly. We can put text blocks on a Model Editor window's background by double-left-

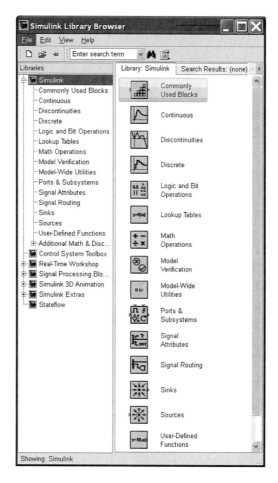

FIGURE 2.1 Simulink Library Browser.

clicking at any point in the background. A text box is opened with a blinking cursor, and we can type documentation into the box. We can use this feature to title models with the model number and a short descriptive title.

If we click on the File menu in the Menu bar of the Model Editor window, we see a list of menu items, including one named Model Properties. Selecting this item opens a Model Properties window containing several tabs. Selecting the History or Description tabs enables us to document the model within the model's properties. The Description tab contains a text block for a complete description of the model and its purpose. The History tab contains a Model History block that contains history lines used to track changes in the model. It is very convenient to have an automatic means of reminding us for a history line every time we make a change, and the

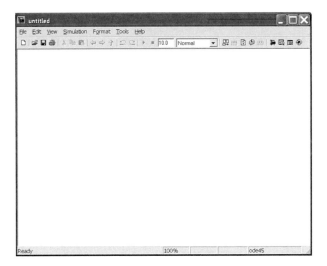

FIGURE 2.2   Simulink new Model Editor window.

History tab has an option at the bottom enabling us to get a prompt each time we save the model. If we select the "When saving model" option at the "Prompt to update model history" item at the bottom, we'll be prompted to enter a history line in the Log Change window for the model. Be sure that the "Show this dialog next time when save" and the "Include modified comments in modified history" options are checked. If we make a trivial change that does not need to be documented, clearing the "Include modified comments in modified history" option allows the model to be saved without a history line added to the history.

Note that the title bar at the very top of the Model Editor window now says "untitled." The word *untitled* is used to indicate that we have not yet named the model. Whenever the symbol * is appended to the model name, it shows that there have been changes to the model since it was opened. At the bottom of the Model Editor window there is an information line showing a status on the left, the amount of magnification (zoom) in the middle, and execution configuration information on the right.

Since it's always a wise strategy to save models frequently, we save the model now. If we select the File | Save As menu item when we are in the Model Editor window, we can use a Windows File Browser to locate the target folder and save our model under a name we choose. The file name will be the name of the model, and it will have the extension .mdl in the File Browser to indicate that it's a Simulink model file. An additional piece of advice is to append a version identifier to the file name so that we can retain different versions as we develop the model. For example, we're going to use Model_chapternumber_modelnumber_version as the file name, so that the first file to be saved is Model_2_1_1, and so on. We might want to use a blank separator or decimal point in our name, such as Model 2.1.1, but Simulink doesn't allow blanks or decimal points in model file names, so we chose to use the "underscore" symbol as a separator.

2.3 SIMULATION OF THE SIMPLEST MODEL

Let's begin our simulation by titling the new model using the method discussed in Section 2.2. We assign the name Model_2_1 for our model filename, create a title text in the empty Model Editor window displaying "Model_2_1. The Simplest Model," and update the model history to show that our version is 1 and that we are making the initial model. Our Model Editor window should look like the one in Figure 2.3.

Next, we turn to entering the most important blocks into the model: the output blocks. These are the blocks that display the results of the simulation to the user, so they should have the primary place in the diagram. We need to create them, place them on the right-hand side of the simulation model, and label them by their names, $x(t)$, $y(t)$, and $z(t)$. Our task at this point is to find blocks in a Simulink Library that provide us with the output blocks.

2.3.1 Output Blocks from the Sinks Library

Turning to the Library Browser, we see the Sinks library. This is the library that contains the blocks used for output. Opening this library, we see a variety of output blocks useful for different output functions.

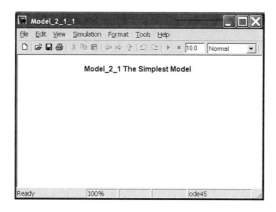

FIGURE 2.3 Model_2_1 for the simplest simulation.

For many years, the *oscilloscope* was the primary laboratory tool for displaying the time behavior of signals. The classical oscilloscope was a cathode-ray tube that presented the viewer with a black background and traced the input signal in time as a white line on a grid of $y(t)$ versus t. So it is not surprising that the basic block for output in Simulink is a simulation of the classical oscilloscope named the Scope block. This output block is very commonly used in simulations created in Simulink, so we use it in this first model.

2.3.1.1 The Scope Block

The block that we need to simulate the output block is the Scope block found in the Sinks library. If we select the Sinks library in the left-hand side of the Library Browser, we see the blocks from this library appearing in the right-hand side, and we can select the Scope block in the list, as shown in Figure 2.4.

To insert a copy of this block into our new model, we just select the block and drag it into the Model Editor window, releasing the mouse button when we reach the place where we want to drop the block. The Model Editor window then looks like the one in Figure 2.5.

We can see that a copy of the Scope block has been created with the default name Scope. The arrowhead on the left of the Scope block is the input connection that is used to provide inputs to the Scope. Next, we need to add two more blocks so that we have one for each output of Equations (2.4)–(2.6). Then we save the model in the folder chosen for models, using the file name Model_2_1_2.mdl.

Before continuing with this simulation construction, let's look more closely at the Scope block output. Double-left-clicking a Scope block will open the Output window, showing what the current output is. It will show a graphical format like the one shown in Figure 2.6, with the value of the input signal displayed on the y-axis of the graph and the simulation time on the x-axis. If no simulation execution has been done, there will be no signal displayed, as seen in Figure 2.6. The displayed range of y-axis values can be set by the simulator, while the displayed time range is set by the simulation length chosen in the simulation configuration settings. Various graph option icons appear on the line above the graph.

The details of the Scope block and its parameters (or any other block) can be found by selecting the block in the Library Browser and clicking the Help | Help for the Selected Block on the Library Browser toolbar.

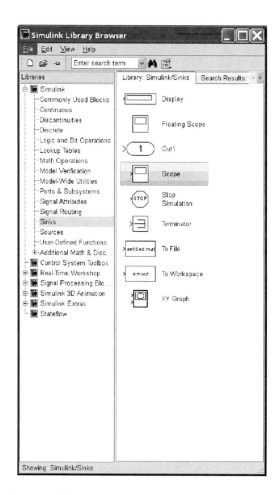

FIGURE 2.4   Selection of the Scope block in the Sinks library.

Simulink provides a comprehensive Help window with information on the meanings of the parameters. An alternative way to see the documentation is to select the block in a Model Editor window and select Help | Blocks on the Editor window toolbar.

Continuing with our model, we want to add names for our three Scope blocks so that we can distinguish them. Each block in Simulink has a Block Name that is printed underneath the block on the Model Editor window. By default, this name is the name of the block type, so that a Scope block has the name Scope underneath it. But the icon for the Scope block is unique, and users should have little problem seeing that it is a Scope block from its appearance, so we can change the name Scope and replace it by a name that

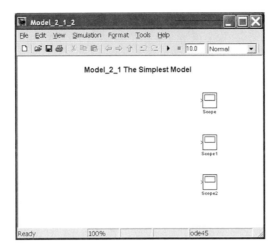

FIGURE 2.5   Model_2_1 with inserted Scope blocks.

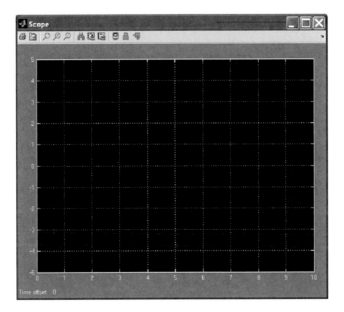

FIGURE 2.6   Example output from the Scope block.

indicates the particular block's purpose or function within the model. One way we can do this is by left-clicking the Block Name and editing the field.

However, a better way of customizing the name is to use the Block Properties window. If we right-click on the Scope block in a Model Editor window, we see the menu of actions on the block. The two actions—Format and Block Properties—are both used for changing the block name.

Selecting the Format action opens the submenu shown in Figure 2.7, containing a number of formatting choices. If we want to hide the default-type name, we only have to select the Hide Name entry in the list.

Next we re-right-click on the Scope block, but this time select the Block Properties action item to see the submenu window below. This submenu has three tabbed panes on it, titled General, Block Annotation, and Callbacks. If we select the Block Annotation tab, we can enter a text string into the "Enter text and tokens for annotation:" text area that will be the name for the block appearing beneath the block in the Model Editor window, as shown in Figure 2.8. If we make these changes to our model, we have our third version of the simplest model, as shown in Figure 2.9.

Now we have to provide the rest of the blocks for the model. We reexamine Equations (2.4)–(2.6) to see which Scope inputs are needed, and we find that we need three constant inputs.

2.3.2 Input Blocks from the Sources Library

The Sources library in the Library Browser contains a variety of blocks for input of data. Opening this library, just as we did with the Sinks library, we see that there is a Constant block available to us.

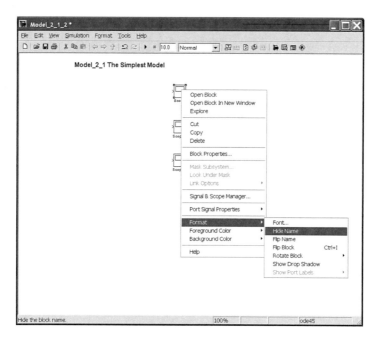

FIGURE 2.7 Hiding the default name of the Scope block.

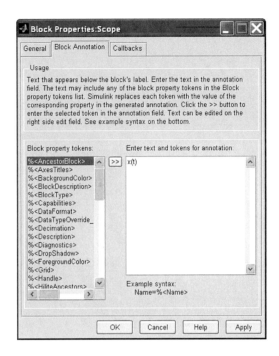

FIGURE 2.8   Providing a user name for the Scope block.

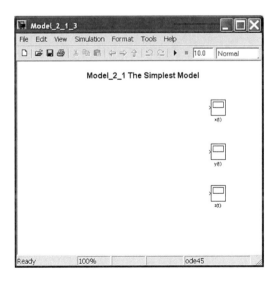

FIGURE 2.9   Model_2_1_3 with named output blocks.

2.3.2.1 The Constant Block

The Constant block, just as its name implies, provides a constant output at all times. All we need to do is to rename the block and change its default value to the value we want. Changing the name of the Constant block can be accomplished exactly as we did with the Scope blocks: We use the Format and Block Properties of the menu displayed by right-clicking the block in the Model Editor window. To change the default value of the Constant block, we can double-left-click the block to open the Constant Block Parameters window, as shown in Figure 2.10.

This window contains a brief description of the block's action at the top, followed by two tabbed panes named Main and Signal Attributes. The line in the Main tab pane named Constant Value can be changed to set the constant to the desired value. Since this is our first encounter with block parameters, it is useful to point out that the default values of parameters are usually the appropriate ones, unless we have a clear understanding of the effects of a parameter change. Changing the value of the Constant is a change that we understand, so it is certainly correct for us to make this change. But it is best to leave the other parameters at their default values unless we know what we are doing. In this textbook, we are studying the basics of using Simulink, so many interesting topics and ways to adapt Simulink blocks through parameter changes cannot be covered. But, after completing this book, it is worthwhile to deepen our knowledge by studying the parameter documentation to see how we can improve on the basic block setups.

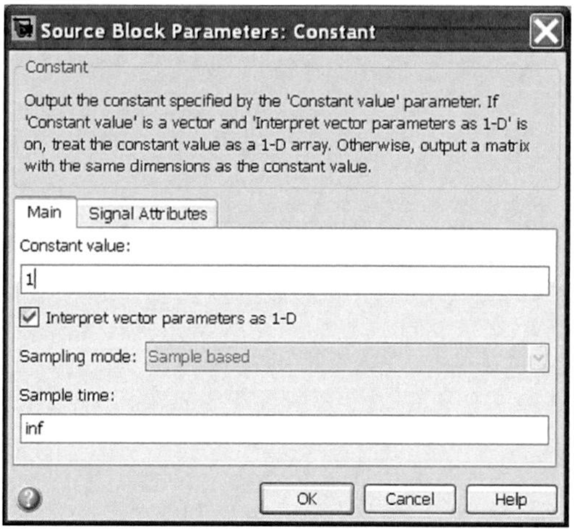

FIGURE 2.10 The Constant block parameters.

Returning to our model construction, we can select this block from the Sources library and drag it to the left-hand side of the model to give us an input block that provides a constant output. We need three of these for the three constant values of the $x(t)$, $y(t)$, and $z(t)$ inputs. Again, we rename all the blocks using appropriate names and produce the model shown in Figure 2.11.

2.3.3 Block Connections

Now we complete our first model by supplying the processing blocks between the input and output blocks. For the simplest model, there is no processing of inputs needed, so we just want to supply the input-block outputs to the appropriate Scope block inputs. This is done by resting the cursor on the output arrowhead of a Constant block, pressing the left mouse button, and dragging the cursor to the input arrowhead of the relevant Scope block. We see that when the cursor rests on the output arrowhead, it changes into single crosshairs, indicating that it is ready to begin a connection starting at that point. When we reach the input arrowhead for another block, the cursor will change into double crosshairs, showing that it can complete the connection at this block. Note also that Simulink has a routing algorithm that is used to create a series of horizontal and vertical legs to draw the route.

If we want a different routing, we can begin our connection as previously stated, but release the left mouse button at a point on the Model

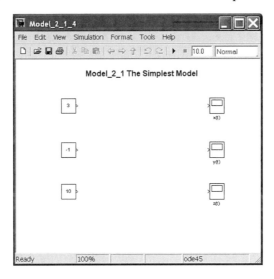

FIGURE 2.11   Model_2_1_4 with input blocks.

Editor window background, where we want the route to be anchored. Then we left-click at the same position and continue this process until we reach the termination arrowhead. A second method of customizing the routing is to make a connection using the Simulink default routing algorithm, and then adjust the route manually afterwards. This is done by resting the cursor on a leg of the route and left-clicking to select the route. The anchor points of the route will appear as black squares. Selecting one of the legs and then dragging it will cause the leg to move vertically or horizontally to create a new route.

If we connect our input and output blocks as discussed previously, Figure 2.12 shows the connected model that results from the connections.

2.3.4 Running the Simulation

We're now ready to execute the simulation. To run a simulation in Simulink, we must first open the Simulation menu in the menu toolbar found at the top of the Model Editor window. Opening this menu, we find a choice named Configuration Parameters.

2.3.4.1 The Configuration Parameters

Selecting this opens the Configuration Parameters window, containing a number of parameters that can be customized. Note that this window has the submenu choices in the pane at the left. For each choice, a pane with appropriate parameters appears on the right. To run our simulation, we need to set some of the Solver parameters. Since this is the default Select choice when the Configuration Parameters window is opened, we see the window shown in Figure 2.13.

There are a large number of parameters that can be set for the Run configuration, but we will follow the strategy of only changing the values that are necessary. First, we need to set the Start time and Stop time. Since this simulation run is demonstrative only, we accept the default values of 0.0 and 10.0. Next, we choose the Solver options. Let's choose the Type to be Fixed-step and the Solver name to be discrete (no continuous states). Lastly, let's change the Fixed-step size from auto to 0.1. Select OK to apply the choices and exit the Configuration Parameters window

2.3.4.2 Observing the Simulation Output

Returning to the Model Editor window, we select Simulation | Start on the toolbar (or we click the black arrowhead on the toolbar) and observe that the status report in the lower-left-hand corner of the Model Editor window

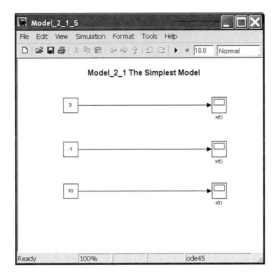

FIGURE 2.12 Model_2_1_5 with connections.

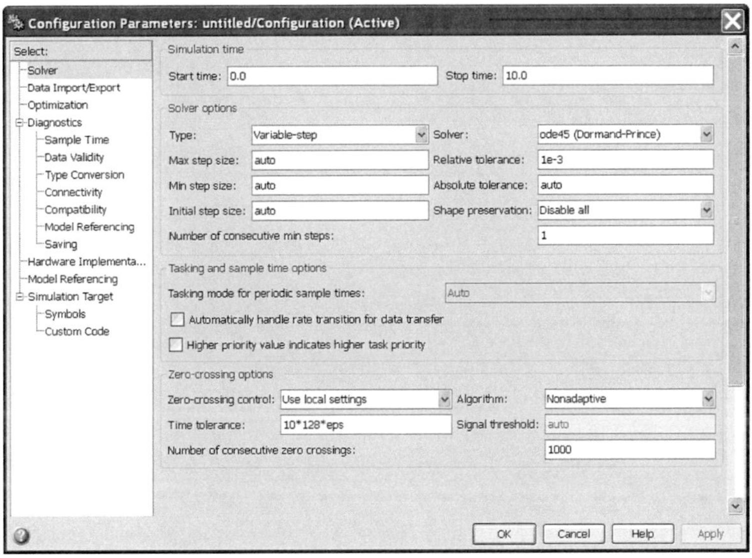

FIGURE 2.13 The Configuration Parameters window.

briefly shows the statuses Compiling, Initializing, Running before changing back to Ready. This shows us that the simulation ran and the results are ready for display. If we double-click the three output blocks named $x(t)$, $y(t)$, and $z(t)$, we see three Scope windows, each displaying the output value on the y-axis and the time on the x-axis. The default range for the output value range is 5 to −5, and the time range is 0 to 10 (as we selected). The $x(t)$ window shows a horizontal line at value 3, the $y(t)$ window a horizontal line at −1, and the $z(t)$ nothing at all. The $x(t)$ and $y(t)$ windows are exactly what we expect, since both of these coordinate values are constant in time, and a horizontal line shows a constant value. The reason we can't see the $z(t)$ line is because it lies outside the default range of the window.

2.3.4.3 Default Ranges and Autoscale

To see the $z(t)$ line quickly, we can click the Autoscale icon on the $z(t)$ window toolbar (the icon is a tiny pair of binoculars), and the window range is adjusted by Simulink to display a range of 9 to 11, with a horizontal line at 10. This is a particularly valuable control if we don't know what value range is expected for the simulation outputs. Using Autoscale on the first run gives us the output range to use in future runs.

If we wanted to rerun the simulation with different constant values for the coordinates of the object, we don't need to close the Output Block windows. We can just change the constant values of the coordinates in the Model Editor window and run the simulation again. The Output Block windows will be rewritten in place by Simulink to display the new results. If the new values are outside the default range, we'll have to select Autoscale again to move the range to encompass the new out-of-range values.

If we need to make several executions of the model, it will be cumbersome to readjust the range using the Autoscale icon each time, so we may want to reset the default range to include all the values expected in any run. This can be done by right-clicking the y-axis of the output window to display a menu containing the Axes properties choice at the bottom. Selecting Axes properties opens a window showing the default range of Y-min and Y-max. These values can be changed, and when the OK button is clicked, the window will show a new default range that remains in force until it is explicitly changed. Note also that the window contains a title parameter that is set by default to the signal name that is being displayed. Since we have not set a name for the input signals yet, there is no title displayed. We will discuss later how to give the signal a name. If we wanted, we could replace the default title value by a text line, and it would be displayed at the top of the graph. Note that the name displayed in the title line for the window itself is still the default block name. This name can be changed only by editing the text box under the Scope showing its default name.

2.3.4.4 Making the Scope Output Printable

The Scope output, as we have seen, is a colored line on a black background. This is a nice presentation for the computer screen, but is not desirable for a printed page. Through the rest of the book, we use a modified copy of the Scope output by processing the captured screen image to change the default colors and to eliminate the surrounding window. Figure 2.14 shows the processed Scope output for $x(t)$. We see that the background is now white, with the output line in black.

2.4 UNDERSTANDING HOW TIME IS HANDLED IN SIMULATION

Simulation time is maintained by the simulation system as a dimensionless variable separate from the real-time clock maintained in the computer. As a result, simulation time can be faster, slower, or the same as real time, depending on our model and the choices we make for the degree of granularity of time in our model. So a one-hour simulation might be accomplished in a second or two on our computer or a one-second simulation in two hours.

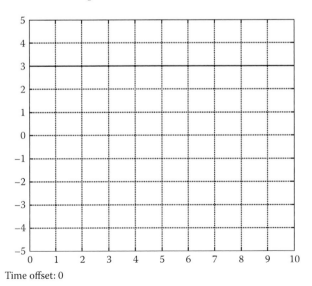

Time offset: 0

FIGURE 2.14 The processed form of the Scope output.

The granularity of time is set by the *time step* that our computer uses to advance the simulation. This time step is the parameter that we set or that Simulink set for us. It can be a fixed value like the 0.01 of the simplest model or a changing value that varies as the simulation unfolds.

In each case, it is important to realize that the units of simulated time are set by us and our model, since the computer just uses dimensionless numbers to execute the simulation. If our model runs with units of seconds, we must remember this so that the x-axis of our output graphs can be interpreted as seconds, rather than some other unit of time.

In the beginning of this text, we will always use a fixed time step so that we can see the effect of our choices. Later on, we will allow Simulink to use a variable time step once we see how this is achieved by Simulink.

2.5 A MODEL WITH TIME AS A VARIABLE

Often we will want to use the time explicitly in our models. To see how we can do this, we now replace our simplest model of a static system with that of a system with very simple dynamics.

Let's consider the system in which the object in a three-dimensional Cartesian coordinate system is moving with a constant velocity. The model for this system will be a set of three equations giving the position of the object in time as functions of the velocity along the three axes and the object's initial position.

$$x(t) = v_0^x t + x_0 \qquad v_0^x = 1 \qquad x_0 = 3 \tag{2.7}$$

$$y(t) = v_0^y t + y_0 \qquad v_0^y = 0.5 \qquad y_0 = -1 \tag{2.8}$$

$$z(t) = v_0^z t + z_0 \qquad v_0^z = -1 \qquad z_0 = 10 \tag{2.9}$$

To create a simulation of this model we proceed exactly as we did in Model_2_1. We open a new Model Editor window, name it Model_2_2, title it, and document it. Now we need to identify the blocks needed to simulate the model of Equations (2.7)–(2.9). Since this model is more complex than our first model, let's break it down to identify the input, processing, and output blocks. Figure 2.15 shows how we can do this. In (a) we list Equation (2.7), reverse it in (b), add grouping brackets in (c) to show the computation precedence of the subexpressions, and show in

(a) $x(t) = v_0^x t + x_0$

(b) $v_0^x t + x_0 \rightarrow x(t)$

(c) $[[v_0^x t] + x_0] \rightarrow x(t)$

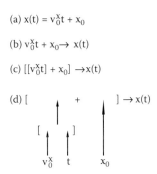

FIGURE 2.15   Transforming the model equations into a block diagram.

(d) how the input data flows into the subexpressions and, ultimately, to the output.

We start by inserting the output blocks to produce version 1. After saving the model, we turn to the input blocks and find that we have seven input blocks: the velocity, time, and initial position for each of the three Cartesian coordinates. The velocity and initial positions can be provided using Constant blocks, just as in Model_2_1. But we also need the simulation time, t, as an input block. The Sources library for Simulink supplies us with a Clock block that provides us with the simulation time.

2.5.1 The Clock Block from the Sources Library

The Clock block is an input block with a single output that provides the value of the current simulation time. There are two parameters we can set to customize this block for Model_2_2. The Display time option allows us to see the current simulation time in a box, since the Clock block icon becomes a rectangle with the simulation time displayed within it. The Decimation parameter allows us to set the redisplay time of the time value within the box to integer multiples of the simulation time step. This is useful if we are running a simulation with a short simulation time step but only need to see box changes at longer time intervals. In Model_2_2 we check the Display time option and set the Decimation to 1, as shown in Figure 2.16, so that the clock display is updated at each simulation time step.

Inserting these seven input blocks produces Model_2_2_2 , as shown in Figure 2.17.

Returning to the question of simulating the model of Equations (2.7)–(2.9), we see that we need to multiply the output of the Clock block by a constant velocity and add the initial position to the result to produce the

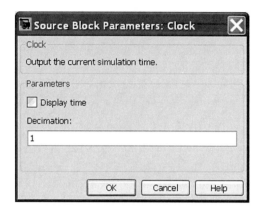

FIGURE 2.16   The Clock block parameters.

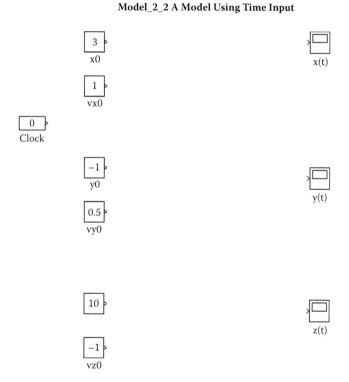

FIGURE 2.17   Model_2_2_2 for uniform linear motion.

desired output for the model. So this model requires processing blocks between the input and output blocks.

2.5.2 Processing Blocks from the Math Operations Library

Now we need to produce the product $v_0^x t$ and the sum $v_0^x t + x_0$ for Equation (2.7) and a similar structure for the other two coordinates. These are both mathematical operations on values of the model, so we look in the Math Operations Library to find blocks for these operations. Since products have higher precedence than additions in mathematical expressions, we construct the product first and then use its output in the addition.

2.5.2.1 The Product Block

The Product block documentation can be found in the Simulink documentation, and it reveals that the Product block can be used with multiple inputs and that each input can be specified as a multiplier or divider input, allowing the block to be used for multiplication and division. The number and type of the inputs is specified on the Product Block Parameters window, shown in Figure 2.18, which can be brought up by double-left-clicking the icon in the Model Editor window.

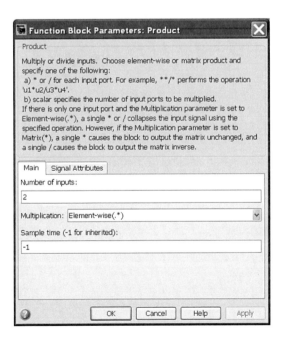

FIGURE 2.18   The Product block parameters.

Again, we have two tabbed panes, with the Main tab as the tab we work with. In the case of a block whose inputs are all multiplied, the line named "Number of inputs:" allows us to specify the number of multiplier inputs by a simple integer. For the case of a block that has multiplier and divider inputs, we can specify the inputs by a list of symbols giving their types, using an asterisk (*) for a multiplier input and a slash (/) for a divider input. We notice that in the second case, the block inputs in the Model Editor window are discriminated by a multiplier or divider symbol.

Returning to Model_2_2, we drag a Product block into the Model Editor window. Since we only need a simple product, we choose the default value for the number of inputs. We also rename the Product block using the product term from the equation as the name. This will help us check our model to make sure that the block diagram is correct. We cannot use sub- or superscripting in the Block Annotation field, so we must place the sub or super terms at the same level.

Now we route the outputs of the Clock block and the relevant v_0 Constant block to the inputs of the relevant Product block. We must now supply the Add block for the last processing block, and we also find it in the Math Operations library.

2.5.2.2 The Add Block

Documentation for the Add block indicates that it is similar to the Product block in the aspect that it may receive multiple inputs and that each input may be designated an adder or a subtracter input. But it differs from the Product block in that a *spacer* (the symbol |) may be inserted into the list of signs to space the input arrowheads on the icon. This is a helpful feature when multi-input addition operations are required, since they can be spaced out for clarity. Note also that there is a second block for addition in the Math Operations library called the Sum block. These two blocks are identical except that the default shape for the Sum block is round, while the default shape of the Add block is square. The circular shape for the operation may be preferred by simulators in fields where the circular shape is commonly used.

If we now drag three Add blocks into the Model Editor window, we can complete the processing block structure. We might rename the Add blocks if it gives clarity to the diagram, but in the case of Model_2_2, the processing structure is so simple that hiding the default name suffices to give us a clean diagram.

We now connect the Product block outputs along with the relevant initial position value to the Add block inputs, and our processing structure is completed. Connecting the Add block outputs to the output Scope blocks completes the entire simulation, which is shown in Figure 2.19.

If we execute this model for a time range of 10, we'll find the output shown in Figure 2.20. The first for $x(t)$ is shown at the top, the output for $y(t)$ in the second, and the last for $z(t)$ at the bottom. This is exactly what we expect from the simulation, since the object moves uniformly through the Cartesian space.

Model_2_2 A Model Using Time Input

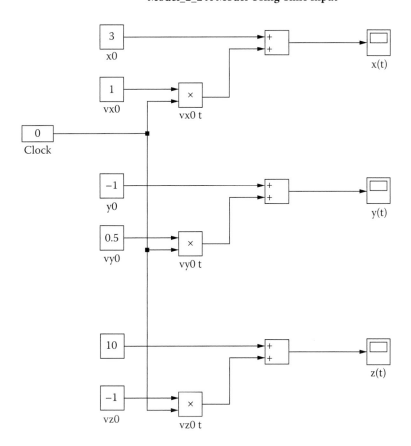

FIGURE 2.19   Model_2_2_3 for uniform linear motion.

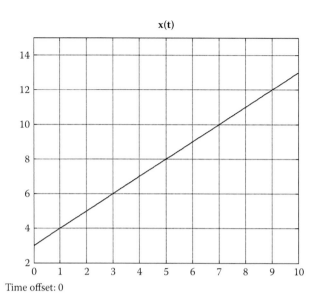

Time offset: 0

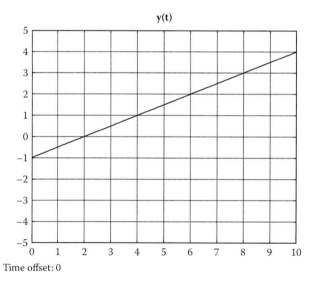

Time offset: 0

FIGURE 2.20 Model_2_2_3 simulation results. The *x*-axes display simulation time, and the *y*-axes display position in arbitrary units. (*Continued*)

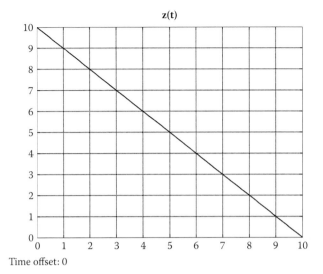

Time offset: 0

FIGURE 2.20 (*Continued*) Model_2_2_3 simulation results. The *x*-axes display simulation time, and the *y*-axes display position in arbitrary units.

EXERCISE 2.1

Simulate the dynamics of a system described by the equation

$$y(t) = \frac{4t}{4+0.7t} + \frac{\pi}{t+1} \tag{2.10}$$

over the range [0,10] (recall that brackets are used in mathematics to indicate that the endpoints are part of the range). Use a fixed time step of 0.01 with a discrete solver and π.

2.6 HOW SIMULINK PROPAGATES VALUES IN BLOCK DIAGRAMS

At this point, it is worthwhile to discuss carefully how Simulink computes the output for Model_2_2_3. When a simulation is ready to be run, we must supply Simulink with a start time, a stop time, and a time step (or else let Simulink supply these from defaults). After we start execution, Simulink reads the model and orders the blocks so that it can begin simulation.

Starting from the Scope block, Simulink looks back along the input lines of the Scope block to locate the blocks that must supply the inputs, as we see in Figure 2.21. If the inputting blocks are sources, their output

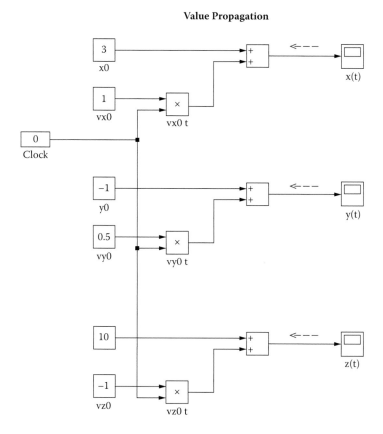

FIGURE 2.21   Simulink value propagation, Step 1.

will be computed using the current simulation time and propagated forward to the Scope block. If the inputting blocks have inputs themselves, Simulink will repeat the same procedure that it has carried out for the Scope block by looking for the inputs to these blocks. During this step for Model_2_2_3, Simulink will find the Add blocks that supply inputs to the Scope block. But it must then repeat the procedure for the inputs of the Add blocks, as shown in Figure 2.22.

Now Simulink finds a Constant block for the initial position, which returns an output immediately, and a Product block, which has a further set of inputs. The Product block must have its inputs evaluated to produce the necessary output, as seen in Figure 2.23.

The inputs to the Product block are traced and end in a Constant block for the velocity component and the Clock block. Both of these

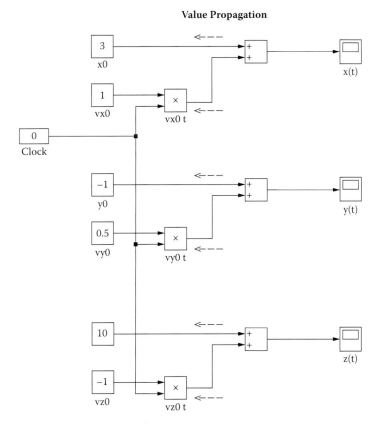

FIGURE 2.22   Simulink value propagation, Step 2.

blocks produce an output without any inputs, so these outputs are propagated forward to produce the Product block output, as seen in Figure 2.24.

The outputs of the Product block are then propagated forward to the Add blocks that now produce their own outputs. Finally, the Add block outputs are propagated to the Scope block and the result is displayed, as seen in Figure 2.25.

Then Simulink advances the simulation time by the next time step and repeats the value propagation process. In this way, Simulink builds up the output over the desired time interval by searching backward along input lines to find values and propagating them forward along the same input lines to produce results.

Value Propagation

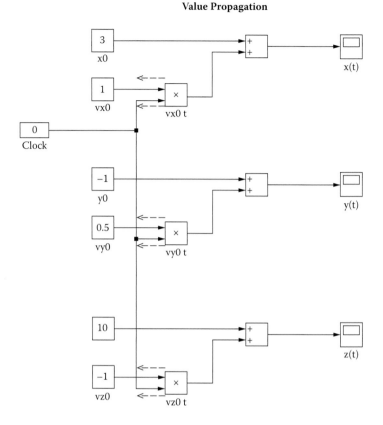

FIGURE 2.23 Simulink value propagation, Step 3.

2.7 A MODEL WITH UNIFORM CIRCULAR MOTION

Let's examine another model of uniform motion that introduces some new processing blocks. If the uniformly moving object is moving in a circle in the x–y plane perpendicular to the z-axis with a speed v_0 and an angular frequency ω_0, its trajectory is a circle. The model for the dynamics of the object is

$$x(t) = -\frac{v_0}{\omega_0}(\cos\omega_0 t - 1) \tag{2.11}$$

$$y(t) = \frac{v_0}{\omega_0}\sin\omega_0 t \tag{2.12}$$

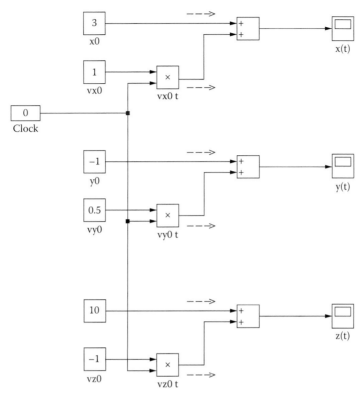

FIGURE 2.24   Simulink value propagation, Step 4.

For our model, we have the initial conditions

$$x(0) = 0 \qquad\qquad y(0) = 0 \qquad\qquad (2.13)$$

a speed of $v_0 = \pi$ m/s, and an angular frequency of $\omega_0 = \pi$ rad/s. Examining Equations (2.11)–(2.12) shows us that there are only two kinds of new blocks needed for this simulation—blocks for the sin $\omega_0 t$ and the cos $\omega_0 t$ terms and a block to negate the v_0 term in Equation (2.11).

Examining the Sources and Math Operations libraries, we find three candidates for the blocks needed for the trigonometric functions: the Sine Wave block from the Sources library and the Sine Wave Function and Trigonometric Function blocks from the Math Operations library. The documentation for these blocks shows that the Sine Wave Function block is used in certain special cases, while the Trigonometric Function block

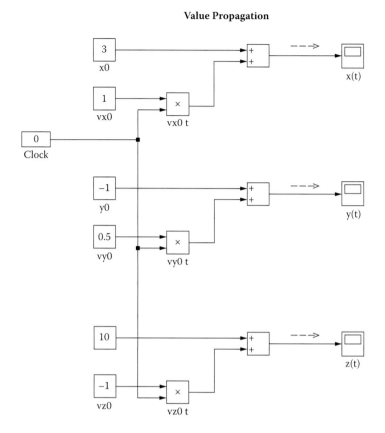

FIGURE 2.25   Simulink value propagation, Step 5.

is more suitable for a term like sin θ than the sin $\omega_0 t$ term that we have, and the Trigonometric Function block would need the external input ωt, requiring additional blocks in the model. So the best block to use here is the Sine Wave block from the Sources library.

2.7.1 The Sine Wave Block from the Sources Library

The Sine Wave block provides the basic function $a \sin(\omega t + \varphi) + b$ for a model. The parameters window, shown in Figure 2.26, allows us to choose all the parameters in the function above to customize the function for our needs. The documentation indicates that the sine type parameter can be time-based with the simulation time as the time (t) parameter. Since our model supplies the amplitude of the sine wave separately, we choose an amplitude of 1. The sine wave is centered at 0, so the bias is set to 0. The frequency of the sine wave term is chosen as π.

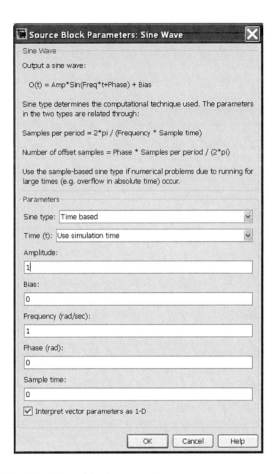

FIGURE 2.26 The Sine Wave block parameters.

Note that choosing a phase of π/2 radians will cause the function

$$a\sin(\omega t + \tfrac{\pi}{2}) + b$$

to be the equivalent of the function $a\cos(\omega t) + b$, so that we can provide either function using the same block. Finally, the Sample time parameter is chosen to be 0, causing the sine wave function to be a continuous output block.

2.7.2 The Gain Block from the Math Operations Library

To produce the negation of the ratio of v_0 and ω_0, we can use the Gain block from the Math Operations Library. This block provides us with a multiplicative constant, and we can enter −1 as its value in the parameters window. The value of the fraction is input to the Gain block and it reverses the sign.

Returning to the construction of the simulation, we construct it exactly as we did the previous two simulations with two Scope blocks for the output blocks $x(t)$ and $y(t)$, and Constant blocks for the input parameters and initial conditions. Figure 2.27 shows the final model of Equations (2.11) and (2.12) named Model_2_3.

We note that, while we could have set the amplitude of the Sine Wave blocks to account for the input parameters v_0 / ω_0, we have chosen to make these inputs an explicitly visible block rather than embed them in the Sine Wave block parameters. This makes our model easier to understand and to change for experimenting with its values.

When we run the simulation, we choose a time range of 10.0, a time step of 0.01, and note that the simulation time units in our model are seconds. Then we get the results shown in Figure 2.28.

These results are what we expect, since the frequency of the circular motion causes one complete period in 2 seconds and the amplitude of the motion is exactly 1. Of course, showing the individual coordinates of the

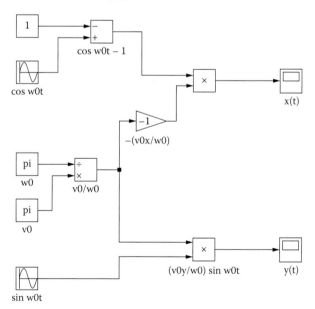

Model_2_3 Uniform Circular Motion

FIGURE 2.27   Model_2_3_1 for uniform circular motion.

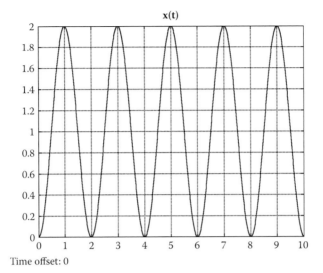

Time offset: 0

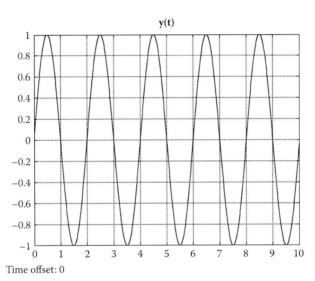

Time offset: 0

FIGURE 2.28 Results for 10-s simulation. The values on the x-axis are seconds, and the values on the y-axis are distances in meters.

object's trajectory using Scope blocks is certainly correct, but it's not the way we would normally show an object moving in a circle in the x–y plane. There is another block, called the XY Graph block, in the Sinks library, that is more useful than the Scope block for showing the trajectory. Let's add this block to Model_2_3_1 to see a better results presentation.

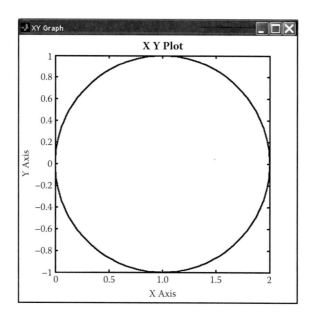

FIGURE 2.29   Final results for a 10-s simulation. The values on the x-axis are the x positions in meters, and the values on the y-axis are the y positions in meters.

2.7.3 The XY Graph Block from the Sinks Library

The XY Graph is a plot of points formed by the two input values, with the upper input used as the x-coordinate and the lower as the y-coordinate at each simulation time step. The range of values displayed is set by entering the maximum and minimum values for each axis in the parameter window for the block. Note that the window is opened when the simulation executes and does not require a double-click to open, as does the Scope block.

Adding an XY Graph block to Model_2_3_1 and connecting both $x(t)$ and $y(t)$ outputs as its inputs provides us with Model_2_3_2. In this final model output, shown in Figure 2.29, we can see the circular trajectory of the object in the x–y plane.

2.8 A MODEL WITH SPIRALING CIRCULAR MOTION

In the last model of this chapter, we use Model_2_3 as our starting point and modify the model to construct a simulation of an object moving uniformly in the x–y plane in circular motion, but spiraling into a smaller circle. The model we are simulating is that of Equations (2.11)–(2.13), modified to have a slowly decreasing radius. The modified equations are

$$x(t) = -\frac{v_0}{\omega_0} e^{-0.1t} (\cos\omega_0 t - 1) \tag{2.14}$$

$$y(t) = \frac{v_0}{\omega_0} e^{-0.1t} \sin\omega_0 t \tag{2.15}$$

with the same initial conditions and parameters as in Model_2_3.

Rather than open a new model and construct all the blocks from an empty model, let's copy Model_2_3 and rename it Model_2_4. We need to update the Model Editor window title as well as the model properties to show the new name.

Examining the new model in Equations (2.14) and (2.15), we see that all we need to change in the model is to add another input to the product of the amplitude and trigonometric function. The new input must be an exponential function of t, and for this we need a new block from the Math Operations library.

2.8.1 The Math Function Block from the Math Operations Library

The Math Function is a very useful block that provides us with a simulation of a number of different functions. It provides 15 different functions consisting of the exponential functions, logarithm functions, power functions, reciprocal functions, and so on. All of these functions take a single input value and provide a single output. Opening the parameter window for a Math Function block and displaying the function list using the down arrow allows the specific function to be picked from a list, as shown in Figure 2.30, which will then cause a new label to be displayed within the icon.

Returning to the completion of Model_2_4, we need to add a Clock block to provide ourselves with the simulation time for input to the exponential function and a Gain block to provide the constant −0.1 multiplier. The outputs of the Gain block produces the input needed for the Math Functions block. Inserting a Math Functions block and choosing the exponential function (this is the default function for the block) allows us to connect the output of the Gain block into the exponential function input.

We now have the desired exponential function, and we can connect it to the product of the initial velocity and radial frequency constants by opening the Product block of these two and changing the number of inputs

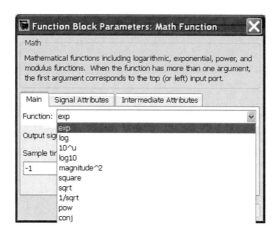

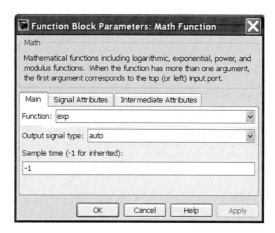

FIGURE 2.30   The Math Functions block parameters.

to three. Reconnecting the inputs then completes the modification of the model to yield Model_2_4_2, shown in Figure 2.31.

If we now run the simulation, we see the results in Figure 2.32 and observe that the coordinates shown do indeed get smaller. The trajectory in the XY graph is now a spiral beginning at (0,0) and ending when the simulation time period of 10 seconds has expired.

EXERCISE 2.2

Simulate the dynamics of the model described by the equation (Shier and Wallenius 2000)

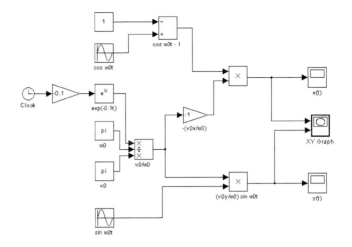

FIGURE 2.31   Model_2_4_2 for spiraling circular motion.

$$y(t) = \frac{2}{\omega^2 - \gamma^2} \cdot \sin\left(\frac{\omega + \gamma}{2} t\right) \cdot \sin\left(\frac{\omega - \gamma}{2} t\right) \qquad (2.16)$$

when $\omega = 2$ and $\gamma = \dfrac{22}{9}$ over the time interval [0,28]. Use a fixed time step of 0.01 with a discrete solver.

2.9 UNCERTAINTY IN NUMBERS AND SIGNIFICANT FIGURES

In the mathematical models we studied in this chapter, all the numbers are pure numbers. By this we mean that the numbers are taken as completely certain numbers, so the results of using these numbers in mathematical operations such as product and addition are viewed as completely certain. If the pure numbers 1.2 and 4.5678 are added, the result is exactly 5.7678, since the pure number 1.2 is equivalent to 1.2000 (or 1.20000000, for that matter). The product $2x$, when the value of x is the pure number 5.25963×10^9, is exactly 10.51926×10^9 or $0.00001051926 \times 10^{15}$ or any other equivalent form. Pure numbers are always taken as completely certain.

But the case is different for mathematical models of physical systems. In these models, the numbers are usually uncertain to some degree, since they result from measurement or calculation from measured numbers. Whenever a number x is measured, the result is a number with a specified uncertainty. Numbers with uncertainly are notated using the terminology

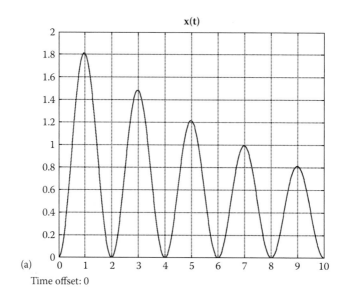

(a)

Time offset: 0

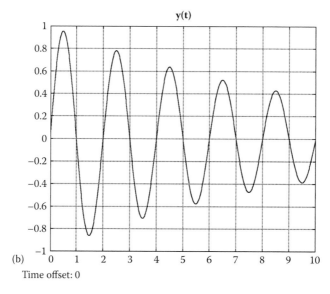

(b)

Time offset: 0

FIGURE 2.32 Results for spiraling circular motion. In (a) and (b), the x-axis displays simulation time in seconds, while the y-axis displays distance in meters. The graph in (c) displays the x–y positions in units of meters on each axis.

(*Continued*)

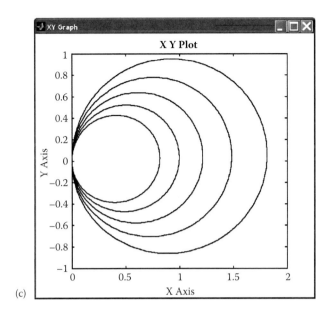

FIGURE 2.32 (*Continued*)   Results for spiraling circular motion.

$x_{\text{best}} \pm \delta x$, where x_{best} is the best estimate of the actual value of x and δx is the uncertainty in the measured value of x (Taylor 1997). There are pure numbers as well in these models, since the integer 2 is a pure number. But it is usually true that most numbers in a model will have some uncertainty in their value.

When we use uncertain numbers in our models, the results obtained from these models are subject to the *propagation of uncertainty*. By this we mean that uncertainties in numbers will propagate through the calculations we use them in, and the results of the calculations will also be uncertain. For example, consider the measured numbers 1.3 ± 0.2 and 4.7 ± 0.3. The product of these numbers, ignoring their uncertainties and treating them as pure numbers, is 6.11. But if we take into account their uncertainties, we see that we could have a result as high as $7.50 = (1.3 + 0.2) \times (4.7 + 0.3)$ or as low as $4.84 = (1.3 - 0.2) \times (4.7 - 0.3)$. While the pure number result is in the same interval as the uncertain number result, [4.84,7.50], the actual center of the interval is 6.17. So the actual accuracy of the result is really only to the first digit in the fraction; the second digit is doubtful.

This problem of accounting for the propagation of uncertainty in calculations has led scientists and engineers to use the concept of *significant figures* and the construction of rules of thumb for their use in calculation.

Significant figures are the digits in a number that are reliably known (Knight, Jones, and Field 2007). So in our example, we have two numbers, each of which has one significant digit (1 and 4) and a product result that has two significant figures (6 and 1). The concept of significant figures is used for determining the number of significant figures in the result of a calculation. For example, one rule of thumb states that the number of significant figures to be retained in a product result is found by rounding the result to the number of significant figures in the less certain of the argument numbers (Taylor 1997). For our example here, the rule would indicate that the result of the product $5.1 \pm 0.01 \times 9.2 \pm 0.01$ would be rounded to two significant figures, which would be $46.92 \rightarrow 47$.

Software implementations of operations do not usually take into account the concept of significant figures, so the results reported by an output block must be suitably modified by the simulator to account for the uncertainties in input parameters. It would be very misleading to a simulation user if result numbers were reported that are far too precise for the input data.

As an example, consider a simple model that predicts an output increasing linearly in time from an initial value of 1.0 (two significant figures) with a slope of 2.4 (two significant figures). The output of this model for 10 seconds is shown in Figure 2.33.

Suppose we want to find the value of the model output when $t = 5.1$. Simulink has a very useful feature of the Scope output that enables us to expand parts of the Scope output easily. If we select a rectangular area within the Scope output, the area will be automatically expanded to fill the entire Scope output. When it does this, the y- and t-axis is expanded to show greater resolution. Figure 2.34 shows the result of expanding the area around the $t = 5.1$ point on the linear output.

As we can see, the value of the output at $t = 5.1$ appears to be closest to 13.25. If we ignore the uncertainty issues, we might be tempted to report a four-significant-figure value of 13.25. But the input data only has certainty of two significant figures, so a better report would be a two-significant-figure result of 13. This is an example of how we must be aware of data limitations in our simulation modeling.

Some problems in scientific computing are known as ill-posed problems because their solution magnifies the uncertainties in the problem because of *loss* of significant figures through mathematical operations (Acton 1996). These and other errors in numerical computation are the subjects of an active research field in advanced applied mathematics and

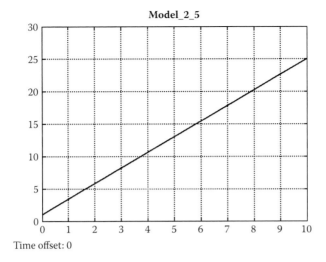

FIGURE 2.33 Significant-figures model output. The *x*-axis displays simulation time, and the *y*-axis displays output, both in arbitrary units.

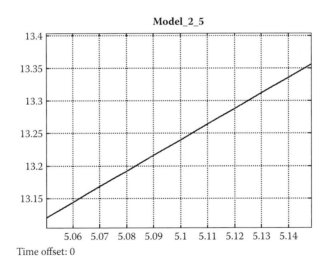

FIGURE 2.34 Significant-figures model expanded output. Arbitrary units on both axes.

science. The references cited here contain a wealth of information on handling uncertain numbers, and the reader should consult them for more details.

2.10 SUMMARY

In this chapter, we covered a number of basics in simulation using Simulink. We have seen how to create and document a new model, how to open the block libraries and select blocks for the model, how to construct the model, and how to run a simulation of the model. We have studied a number of basic output, input, and processing blocks that will be useful in almost all our future models. Our models have been limited to feedforward diagrams in which the values flow only in the forward direction (inputs to outputs).

We have seen that the Library Browser, containing two columnar subwindows showing a list of the block libraries on the left and the blocks in the currently selected library on the right, is the main window for locating blocks to use in our simulation. To insert a copy of a block into a model, we select the block in the library and drag it into the Model Editor window.

We discussed simulation time, which is maintained by the simulation system as a dimensionless variable separate from the real-time clock maintained in the computer. The granularity of time is set by the time step that we specify to advance the simulation.

We studied a number of basic blocks consisting of the Scope block as the fundamental output mechanism; the Constant block, providing a constant value on its output; the Clock block, providing the value of the current simulation time; the Product block and Sum block, providing the product and sum of their inputs, respectively; and the Math Operations library blocks used for simulating mathematical functions.

We saw that Simulink propagates values to output blocks by looking back along the input lines to the input-supplying blocks, until sources are found. Then it propagates the results forward.

We also discussed that simulators must be aware of data-precision limitations and take them into account when reporting simulation results.

REFERENCES AND ADDITIONAL READING

Acton, F. S. 1996. *Real Computing Made Real: Preventing Errors in Scientific and Engineering Calculations*. Princeton, NJ: Princeton University Press.
Knight, R. D., B. Jones, and S. Field. 2007. *College Physics: A Strategic Approach*. New York: Pearson Addison Wesley.

Moler, C. B. 2004. *Numerical Computing with MATLAB*. Philadelphia: Society for Industrial and Applied Mathematics.

Shier, D., and K. Wallenius. 2000. *Applied Mathematical Modeling: A Multidisciplinary Approach*. Boca Raton, FL: Chapman & Hall/CRC.

Taylor, J. R. 1997. *An Introduction to Error Analysis: The Study of Uncertainties in Physical Measurement*. Sausalito, CA: University Science Books.

The MathWorks, Inc. 2007. http://www.mathworks.com.

CHAPTER 3

Simulation of First-Order Difference Equation Models

Iᴺ ᴛʜɪs ᴄʜᴀᴘᴛᴇʀ, ᴡᴇ are going to study systems whose behavior can be described by giving a sequence of values for the dynamic variables or properties. These values are supplied at regularly spaced points on the independent-variable line, and, since we are mostly interested in time-dependent systems, we want the values at times separated by a constant time interval.

Systems of this kind are often found in everyday life. When ecologists or biologists want to track the change in population of an animal species, they may visit a representative site at regular intervals to count or estimate the population. Intervals of a year give a sequence of values that represent the annual changes in population.

Of course, scientists want to do more than just record data. They want to construct a model that explains the relationships between the annual population values and important parameters of the surrounding environment. If it is only necessary to model the system at finite intervals, this model will often be described in the form of dynamical equations called *difference equations.*

3.1 WHAT IS A DIFFERENCE EQUATION?

A difference equation is an equation that relates the value of some fundamental system parameter or function to its past values at earlier times. If the system function we want to simulate is $y(t)$, and we are only interested in its values at times $t_n, t_{n-1}, \ldots, t_0$ (t_0 is the earliest time), then we can use a difference equation to relate the value at t_n to the values at $t_{n-1}, t_{n-2}, \ldots, t_0$. To make our equations easier to understand, we use the notation where $y_n = y(t_n)$, $y_{n-1} = y(t_{n-1}), \ldots$, and $y_0 = y(t_0)$. Then we write the general definition below as it would be found in a mathematical text.

> A difference equation has the form $y_n = g(y_n, y_{n-1}, \ldots, y_0, n)$ for a specified range of values of n.

In this definition, we see that any of the earlier values, the current value, and the time interval number can be combined in an arbitrary function to provide the desired value.

3.1.1 Difference Equation Terminology

Before we go further, let us consider some terminology that is commonly used. A difference equation is called *linear* if the relationship only has terms that are linear in y_i. As an example of a linear difference equation, consider the following:

$$y_n = 3y_{n-1} - 2y_{n-2} + n \tag{3.1}$$

The terms in this equality all have the y_i in the first power, and there are no terms involving products of y_i.

The *order* of a difference equation tells the reader how many preceding time values that the value of y_n may be dependent on. In Equation (3.1), the value of any y_n depends only on the two immediately earlier times, so y_n obeys a *second-order* difference equation.

Knowing the form of the difference equation, however, is not enough for us to determine the value at an arbitrary point in the range. To do this, we also need to know certain *initial values* for the system. For our Equation (3.1), we must have two initial values, since we can't compute the $n = 2$ term without knowing the values at $n = 1$ and $n = 0$.

For Equation (3.1), we could have a completely specified, linear, second-order difference equation with

$$y_n = 3y_{n-1} - 2y_{n-2} + n \qquad\qquad n \geq 2, \quad y_0 = 2, \quad y_1 = 1 \qquad (3.2)$$

From the discussion above, it is clear that we could solve this equation for any desired value of y_n by recursively computing a value of y_i at each period beginning with $n = 2$.

$$y_2 = 3 \cdot 1 - 2 \cdot 2 + 2 = 1 \qquad\qquad (3.3)$$

$$y_3 = 3 \cdot 1 - 2 \cdot 1 + 3 = 4 \qquad\qquad (3.4)$$

and so on.

Difference equations are very familiar in computer science and mathematics, since they can be viewed as recursion equations, and there are well-known methods for solving certain kinds of these equations (Gersting 2003). But in this text, we are only interested in simulating such equations.

3.2 EXAMPLES OF SYSTEMS WITH DIFFERENCE EQUATION MODELS

Keen and Spain (1992) describe a classic experiment in the dynamics of flour beetle populations. The experiment is done by placing a few adults into a suitably prepared container of flour. Every 30 days, the flour is sifted to allow the biologists to count the populations of adult beetles, larvae, and pupae. This data enables scientists to show the change in population based on 30-day time intervals. This system can be modeled by a difference equation model.

Levy (2004) describes the use of finite difference models for solving a model of asset pricing in computational finance. An option represents an asset with a changing value, and the change in its value is described by the Black-Scholes equation under different call schemes. This partial differential equation can be cast into a finite difference model with some definite advantages over other solution techniques. The finite difference model is used to determine the value of an option from the option maturity time (the time at which the option must be exercised or lose its value) to the present, enabling the simulator to choose between different options.

3.3 FIRST-ORDER DIFFERENCE EQUATION SIMULATION

Let us now examine how we can create a Simulink® simulation of a system with a first-order difference equation model. We'll use the simple difference equation model in Equation (3.5) to see how to construct a simulation. Since this is a pure mathematical example, we regard all numerical values as completely precise.

$$y_n = 0.5y_{n-1} + n \quad n \geq 1, y_0 = 1 \tag{3.5}$$

Remembering that the default order of Simulink models is left-to-right (see Chapter 2, Section 2.1), let's reverse our model equation and change the equal sign to an arrow so that the equation will be in the identical order as our block diagram.

$$0.5y_{n-1} + n \rightarrow y_n \tag{3.6}$$

The desired output of the simulation model is the value of y_n, so let's start our simulation by inserting a Scope block at the right to display this value. The input to the Scope block will be the left-hand side of Equation (3.6). We also suppress the name of the Scope block because its unique icon clearly identifies it as a Scope, and we set the Block annotation to "Yn Display." Review Chapter 2 for a refresher of this technique.

Next, we need to add a network of blocks to provide the input to the Scope block. The left-hand side of Equation (3.6) shows us what the structure of the network must be. It must be a sum of two inputs, each input representing one of the two terms in the sum. In Chapter 2, we saw how to create an Add block of inputs, so our next version is the addition of an Add block, connecting its output to the input of the Scope block and increasing its size by using the corner handles, if desirable. Recall that we can also use the Sum block for this. In Chapter 2, we used the block annotation parameter to rename our blocks, but in this model we use a different means of naming: the output ports.

3.3.1 The Input and Output Ports of a Block

If we select a block in a model window, right-clicking produces the properties list for the block (the same list where the Block Properties item appears). One of these items is Port Signal Properties. Selecting this item displays a list of two signal port types — the input ports and the output

FIGURE 3.1   Signal properties for an output port.

ports. If the ports on the block are connected, we see a list of ports and can select one of these for customization. Selecting one produces the Signal Properties window for that port, as shown in Figure 3.1.

At the top of this window, we see that the Signal name can be assigned a test value, and, after we have assigned it a value, the line corresponding to the port is updated to display a label. This is a convenient way to document the flows throughout the model and does not occupy as much space as a block annotation does. Note that port numbers are assigned from top to bottom for a block, so blocks with multiple inputs or outputs have their identities differentiated by this number.

Returning to the naming of the Add block, let's rename its output port as "Yn" so that it is clear what the output value represents. Now we add the two terms in the sum to the network as inputs to the Sum block. The second term can be provided by just inserting the Clock block that we used in Chapter 2 to provide the simulation time interval. We add this block to the first input and change its output port name to "n," with the result shown in Figure 3.2.

Before continuing, this is a good time to make sure that the simulation solver is set to a Fixed-Step solver and that the step size is equal to

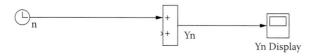

FIGURE 3.2 Model_3_1_1 of the simple, first-order difference equation simulation.

1. Open up the Configuration Parameters screen, as we did in Chapter 2, and set the solver option Type to "fixed-step," the Solver to "discrete," and the Fixed-step size to 1.0, as shown in Figure 3.3. Note that the fixed-step size must be 1.0, since our difference equation model is only defined for integral values of n.

Now we add the two terms that are input to the Sum block. The first term, $0.5y_{n-1}$, is a product of two inputs, which we discussed in Chapter 2, and this is easily added to our next version by using a Product block and a Constant block. Remember to set the value of the Constant block to 0.5 using the Constant Parameters window. Now we need to add the second of the two inputs to the Product block, y_{n-1}, and this causes us to investigate two new blocks: the Memory block and the IC (Initial Condition)

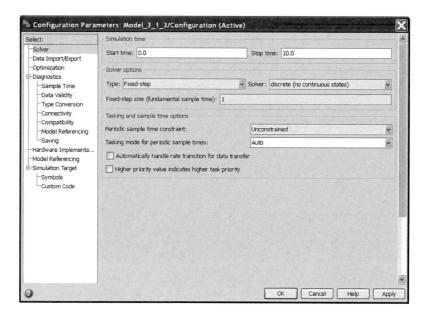

FIGURE 3.3 Configuration parameters for the simple, first-order difference equation simulation.

block. We need the Memory block to retain the last value we computed for y_i, that is, y_{n-1}, so that it is available in the next time step. The IC block is needed to provide a way of starting the y_i values with the correct starting value for y_0.

3.3.2 The Memory Block from the Discrete Library

The Memory block is a block from the Discrete library, and it has the behavior of retaining its input from the previous time step for outputting at the next time step. So at each time step, two things happen at the Memory block.

1. The value stored in the block is output.

2. The input value is stored in the block, destroying the value used in step 1.

From this description, it is easy to see why it is called a Memory block; it remembers the last value and acts exactly like a computer memory element. The Memory parameters can be accessed by double-clicking the block. But we will use the default values of the parameters for now.

Returning to Model_3_1, this is exactly what we need to simulate the y_{n-1} term during the time step when we compute the y_n value. But where do we get the last value to store in the Memory block for the next time step? The answer is that we get it from the output of the entire network at the previous time step! Figure 3.4 shows the product box connection that provides the input for our Memory block.

This is our first encounter with a *feedback loop*. The signal path in a feedback loop goes from output back to input; it goes right-to-left instead of the left-to-right order we have used so far.

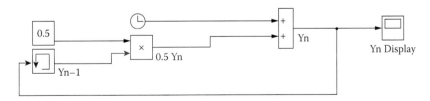

FIGURE 3.4 Model_3_1_2 of the simple, first-order difference equation simulation.

3.3.3 The Feedback Block Diagram

In modeling physical systems, we often find that signals generated on the right-hand side of the block diagram must pass backwards to the left-hand side in opposition to the flow of the feedforward diagram. This kind of diagram is called a *feedback diagram*, since some values flow in the backward direction (outputs to inputs). Feedback diagrams are conceptually difficult when they are first encountered, and finding the correct point at which to reconnect the feedback into the diagram requires a fine judgment. Actually constructing them with our simulation software, however, is not much more difficult than constructing feedforward diagrams.

Feedback is the capability in our simulator that enables us to simulate a wide variety of systems that need an input signal for control or other purposes. The importance and power of this concept cannot be overstated. It is vital to all our modern systems by making possible a variety of behaviors that would be unachievable otherwise. Systems ranging from the ordinary thermostat to the orbiting satellite use this idea to make their behavior stable and responsive to environmental changes. It is an impressive achievement of the human intellect that this concept was developed and is used on a routine basis.

When we first see this kind of feedback network, it looks like it won't work. How can it be that the output value is determined by itself? But the magic comes in the storage of the signal in the Memory block. Each time a computation of the output value is made, only the stored value is used in the new output computation, not the incoming value. And this is another important concept to grasp. The stored value is called the *state* of the Memory block.

The state of a system encodes the history of its inputs and outputs so that the system can use its history together with the current inputs to determine its current response

---

In science, we are concerned with creating functions that produce some desired output behavior. But how can we create a function whose behavior depends on its past as well as its current inputs? The answer is the *state function*. This kind of function has, as part of its computational capability, a state that supplies the previous values of the function so that they can be used along with the current inputs to provide a new output.

Many of the most familiar functions from mathematics are not state functions. For example, the square function has an output that depends only on its current inputs, not on the previous values of the square function. So when we write $y = x^2$, we only need to know the current value of x, not any previous values of y.

But our difference equation example *is* a state function. We must know the previous value of y, along with the current time step n, to compute the next value of y. And many naturally occurring systems as well as most artificial systems are of this kind. Thus it is necessary to provide our simulations with a block to perform this function.

The Memory block provides the simplest state-dependent function. Its state is composed of only one previous value, the last input, and the block output does not depend on any other inputs, so its output is determined entirely by its state value.

Now in our model, the Memory block holds the last value for the next computation. But we have one more problem to solve. How do we start the simulation with the correct initial conditions?

3.3.4 The IC Block from the Signal Attributes Library

What we would like to have is a block that produces the correct initial conditions on output on the first time step, but whose output is no longer used for the remaining time steps. Simulink provides us with this block in the Signal Attributes library, and it is called the IC (Initial Condition) block. The behavior of an IC block is that it produces an output at the first time step and disappears from subsequent time steps. It's as if it were taken out of the model for all time steps after the first time step.

To use this block, we want to insert it into the model in such a way that it causes the simulation to compute the correct initial value and also sets the state for the memory to the correct value for the next time step. The place for this insertion is between the Memory block and the Product block so that the y_{n-1} term part of the simulation produces the correct value at $n = 1$. The range of values for the simulation is $n \geq 1$, so we want the simulation to begin at $n = 1$. Now we need to set the IC parameters.

Double-clicking the IC block opens the IC Parameters window, as shown in Figure 3.5. There is only one parameter to set—the initial value. Entering the desired initial value 1.0 into the field completes our setup of the block. This gives us the final model in Figure 3.6.

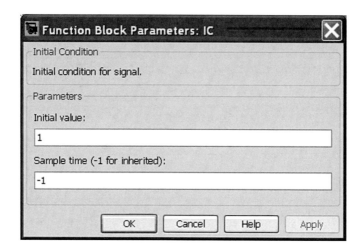

FIGURE 3.5   The IC block parameters.

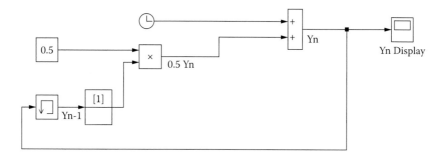

FIGURE 3.6   Model_3_1_3 of the simple, first-order difference equation simulation.

Before we run the simulation, we must also set the starting time for the solver to time step 1, not 0 (the default starting time), since the range is $n \geq 1$. We can do this by opening the Configuration Parameters and setting the start time to 1.0, rather than 0.0 (see Figure 3.7). We'll also arbitrarily pick 10.0 as our stopping time. Finally, we need to set the step size to 1.0.

Now we are ready to run our simulation. If we open the Scope display screen and run the simulation, we will see the output shown in Figure 3.8. Note that Simulink may start the time-axis label of the output plot at 0, although the value shown is the value for an initial time step of 1.0. When it does this, Simulink informs us that there is a time offset of 1.0, using the status message in the bottom left-hand corner of the screen. So we must mentally correct the time axis to the range 1.0 to 10.0.

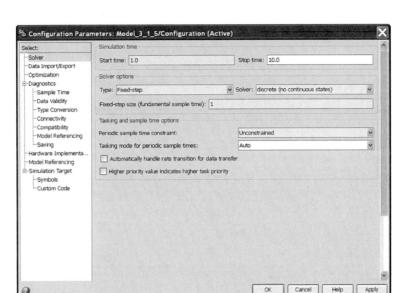

FIGURE 3.7 Correcting the start time for the difference equation range specification.

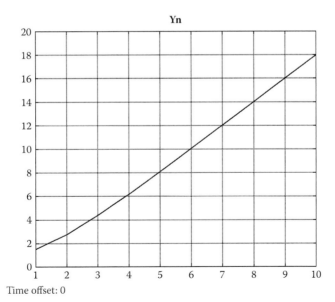

FIGURE 3.8 Output for the simple, first-order difference equation simulation. Simulation time is shown on the x-axis, and simulation output is shown on the y-axis in arbitrary units.

TABLE 3.1   Output Result
Values of the First-Order
Difference Equation

n	y_n
1	1.5
2	2.75
3	4.375
4	6.1875
5	8.0938
6	10.047
7	12.023
8	14.012
9	16.006
10	18.003

The actual values computed during the execution are shown in Table 3.1. If we want to verify the simulation values, it is easy to use a simple spreadsheet to verify them.

3.3.5 Setting Initial Conditions with the Initial Conditions Field of the Memory Block

A different way of providing initial conditions for the example is to use the built-in initial-condition field of the Memory block parameters and forgo adding the IC block to the model. If the Memory Parameters window is opened for the Memory block in Model_3_1_3 of our example, we see Figure 3.9.

Note that the screen allows us to set an initial condition for the output of the Memory block. If we set a value of 1 in the field, the memory block will output a value of 1 at its output port at time step $n = 1$. Then the value 1 would be used along with the constant 0.5 and the time step 0 to compute a y_0 value of 1.5. If we make this change, i.e., delete the IC block from the model, and run the simulation, we see an output identical to that shown Figure 3.8.

The first method (the use of an IC block) is preferable to the second method (the use of the Memory block initial-condition parameter), although it does increase the number of blocks in the simulation. The main reason for this is that the inclusion of an IC block makes the location of the initial conditions explicit to a person viewing the model. This is very helpful to those who have been given the job of maintaining a simulation, since they can orient themselves to the model quickly. When

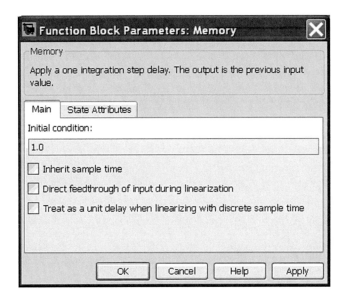

FIGURE 3.9   Setting the initial condition for the Memory block.

important parameters of a model are buried within internal parameter fields, the job of locating them is made more difficult. It also makes the presentation of the model to a nonexpert audience easier, since the initial conditions can be pointed out in an image of the model, without requiring an online version of the simulation. Throughout this book, we use the IC block method for visual inspection clarity.

EXERCISE 3.1

Simulate the dynamics of a system described by the difference equation

$$y_n = 0.2 y_{n-1} - (1 - e^{-n})n \qquad n \geq 1, \quad y_0 = 1 \qquad (3.7)$$

over the range [1,10] (recall that brackets are used in mathematics to indicate that the endpoints are part of the range). Assume that all numerical values are completely precise.

EXERCISE 3.2

A simple economic model for the relationship between supply and demand can be given with the difference equation

$$p_{n+1} = a - \frac{b}{k} p_n \qquad\qquad (3.8)$$

where p_n is the price at time n.

Create a Simulink simulation that shows the change in price for 20 days if the initial price is $p_0 = 30$, the constant $a = 50$, and b/k is a constant. Run the simulation for three values of b/k: 0.9, 1.0, and 1.1. Assume that each day's price remains constant throughout the day and changes overnight. Since this is a physical system, pay attention to the number of significant figures in your result.

EXERCISE 3.3

Bernstein (2003) describes a model for the population of ground squirrels that breed only in the spring, with the population growth of ground squirrels in an area given by the discrete model

$$P_n = R \cdot P_{n-1} \qquad n \geq 1 \qquad\qquad (3.9)$$

where P_n is the number of squirrels in the nth year, and R is the geometric growth factor given by

$$R = 1 + (B - D) \qquad\qquad (3.10)$$

where B is the number of births per individual per year, and D is the number of deaths per individual per year.

1. Simulate the 10-year population growth of an initial population $P_0 = 100$ squirrels with $B = 0.06$ and $D = 0.01$. Once again, make sure that your results reflect the precision of the data.
2. Find the birth rate B that would cause the population to double in 10 years.

3.4 EXAMINING THE INTERNALS OF A SIMULATION

As we develop more complex models, we find that we need to do more than just examine the inputs and outputs of blocks to debug our simulation. We could do this by inserting Scope blocks at various internal points, connecting the test points to the Scope inputs, and using the Scope display to see the intermediate values. But this is time-consuming and tedious, since we have to put them into the model one by one and then remove them when debugging is finished. It would be much easier if we could use one Scope block, and display the values at the desired points without physically connecting them to the Scope. Simulink provides us with just the block we need.

3.4.1 The Floating Scope Block from the Sinks Library

The Floating Scope block is a kind of Scope block that allows us to use a tabular display to pick the points from which the inputs are taken. So we don't need to connect the Floating Scope physically to the points in the model. Let's use our simple, first-order difference equation model to see how this block is used.

First, we open our previous model, Model_3_1_3, and insert a Floating Scope from the Sinks library into the model at the bottom. Do not try to connect any input lines to it. Let's also leave the default block name on the Floating Scope block so we can differentiate it from the regular Scope block. This produces the model shown in Figure 3.10.

Now we need to pick the signals displayed by the Floating Scope. The easiest way to choose the connections is to double-click on the Floating Scope block to open up the display window for the block. Then left-click the rightmost icon on the Toolbar; it should show the hint "Signal Selection." A new window will open up, showing the signal selection editor. Both of these windows are shown in Figure 3.11.

Note that the signal menu in the right panel of the Signal Selector shows the signals that can be displayed in the Floating Scope. Let's select Yn (output of the Sum block), n (output of the Clock block), 0.5Yn (output of the

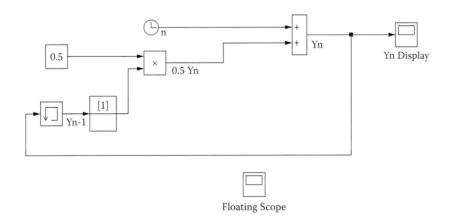

FIGURE 3.10 The difference equation model with a Floating Scope block.

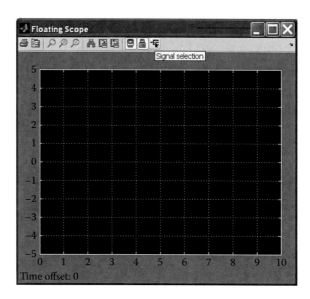

FIGURE 3.11A Signal selection for a Floating Scope block.

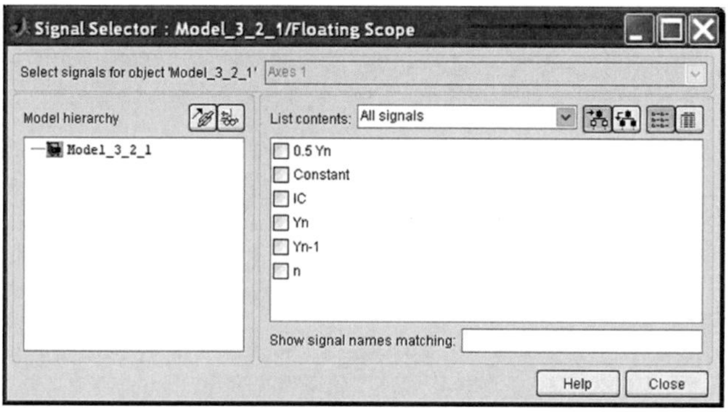

FIGURE 3.11B Signal Selector Model_3_2_1/Floating Scope.

Product block), Yn-1 (output of the Memory block), and the Constant block output. We check their boxes and close the Signal Selector.

Before running the model with the Floating Scope, we need to change one Configuration Parameters item. If we select Simulation on the menu bar and then choose Configuration Parameters | Optimization, we see a number of options. If the Signal Storage Reuse item is checked, uncheck it and close the parameters selection. We'll return to a discussion of this configuration parameter later in the text.

If we run the model, we'll see the output from the Floating Scope as shown in Figure 3.12. On the computer screen, Simulink uses different line styles for each signal being displayed in the Floating Scope so that we can tell which signal is which. The first assignment starts with the first block created in the model and proceeds serially to the last block created. The order of the blocks can be determined by running the Debugger and selecting the Sorted List tab. There are four line styles used, and their selection for use proceeds in the order shown in Table 3.2.

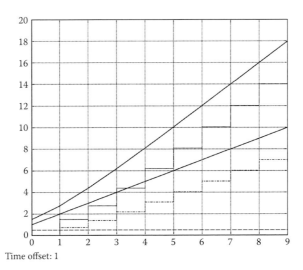

Time offset: 1

FIGURE 3.12 The multiple signal output from the Floating Scope block. Simulation time is shown on the x-axis, and simulation output is shown on the y-axis in arbitrary units.

TABLE 3.2   The Order of Line Style Used in the Floating Scope

Line 1	Solid
Line 2	Dashed
Line 3	Dotted
Line 4	Dash-Dotted
Line 5	Solid
Line 6	Dashed
and so on	and so on

TABLE 3.3 The Order of
the Blocks in Model_3_2

0:0	Clock
0:1	Constant
0:2	Memory
0:3	IC
0:4	Product
0:5	Sum
0:6	Scope
0:7	Floating Scope

Using the Debugger (see Appendix C on how to use the Debugger), we find that the order of the blocks is as shown in Table 3.3.

So, in Figure 3.12, we have the output of the Clock, n, shown with a solid line; the Constant output, 0.5, shown with a dashed line; the Memory output, Y_{n-1}, with a dotted line; the Product, $0.5Y_n$, with a dash-dotted line; and the Sum output, Y_n, with a solid line.

EXERCISE 3.3

Add a Floating Scope to the model for Exercise 3.2, the population growth of ground squirrels, and display the value of R throughout the time-step range of the simulation. Verify that it remains constant in this system.

3.5 ORGANIZING THE INTERNAL STRUCTURE OF A SIMULATION

Models can become very complex if we create them as unstructured, flat diagrams. Consider how we might simulate a system with the following difference equation model, where all the numbers are completely precise:

$$y_n = 2y_{n-1} + 3(\sin n\pi)y_{n-2} + 4n^2 y_{n-3} + 5(2 + e^{\frac{n}{2}}) \tag{3.11}$$

for $n = 3, \ldots, 6$ with $y_0 = 1$, $y_1 = 2$, and $y_2 = 3$.

Fundamentally, this equation is no more difficult to simulate than Equation (3.5), but it is more complex in its term compositions, so its simulation requires more blocks and combinations. In Figure 3.13, a possible model is shown. This simulation is complex to analyze visually or to modify and extend.

We would like to *abstract* the details of this simulation so that we can visually inspect the model more easily and modify and extend it with minimum labor.

Abstraction is the process by which humans suppress irrelevant details of a system in order to focus on its salient features.

The advantage of doing this is that it simplifies the system structure into its essentials and allows us to work only on the important parts of the system. This facility is another powerful technique by which humans amplify their ability to deal with complicated and complex problems.

Simulink provides us with an abstraction mechanism through the user-defined Subsystem block.

3.5.1 The Subsystem Block from the Ports and Subsystems Library

A user-defined subsystem is a combination of elements, selected by the user, that have inputs and outputs. It is a mechanism for encapsulating unimportant complexity into abstractions so that we can reduce the visual complexity of our simulation model. We can examine the use of this mechanism by encapsulating the layers of the simulation in Figure 3.13 into abstractions representing the right-hand-side terms in Equation (3.11).

To create the subsystem, we select a structure of blocks and connections with a selection box. We do this by positioning the cursor on a location in the background of the model, pressing the left button down and holding it, then moving the cursor to expand the box around all the desired blocks and connections, as shown in Figure 3.14.

Next, we select the Create Subsystem item from the Edit menu. Simulink will automatically turn the selected structure into a subsystem with the correct number of inputs and outputs (in this case, two outputs only). Figure 3.15 shows the resulting diagram.

We will use the default parameters for the Subsystem parameters, so we do not change any parameter values. The subsystem can be expanded into its parts at any time by double-clicking the Subsystem block, as shown in Figure 3.16.

If we look at this expansion, we see that Output port Out1 is the term we were computing in the nonabstracted model and Output port Out2 is just

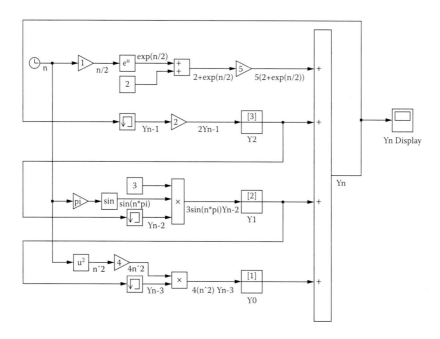

FIGURE 3.13   Simulation of the model in Equation (3.11).

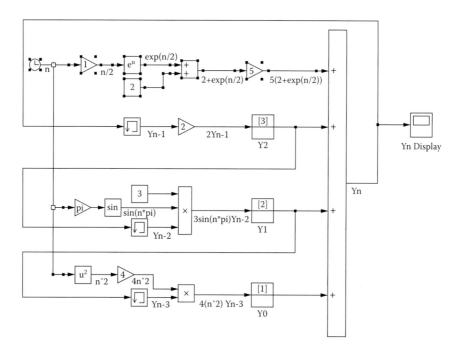

FIGURE 3.14   Selecting the top layer of the model in Figure 3.13.

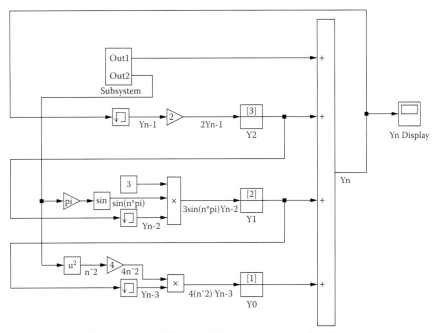

FIGURE 3.15   The top layer of the model in Figure 3.14 after abstraction.

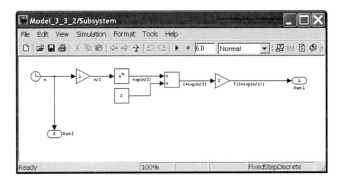

FIGURE 3.16   Expanding the Subsystem block for the top layer of the model.

the value of n that we used in layer 4. Let's rename these so we can identify which signal in our nonabstracted model is which in the abstracted model. But we will leave the block name visible so that it serves to remind us that this block is actually a subsystem and must be expanded to show the details.

Closing the subsystem expansion and renaming the output port properties gives us the model in Figure 3.17.

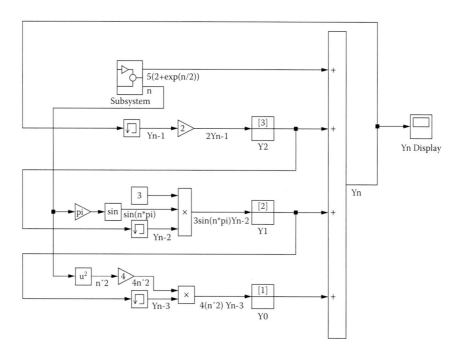

FIGURE 3.17   Renamed output ports of the Subsystem block for the top layer.

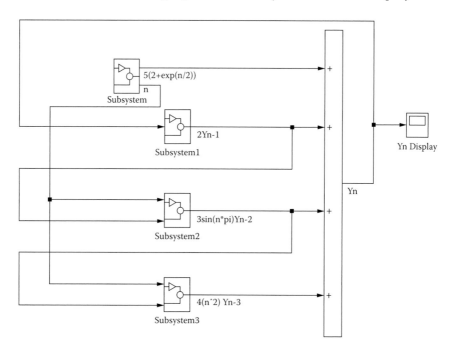

FIGURE 3.18   Final model encapsulating all layers.

We can continue this process by encapsulating the remaining three layers. When we are finished, the final top-level view of the model is shown, as seen in Figure 3.18.

EXERCISE 3.4

Simulate the system described by the first-order difference Equation (3.12), producing a model that has only Subsystems and a Scope in the top-level diagram. The numbers are all precise values.

$$y_n = \frac{1}{2}(\cos n\pi)y_{n-1} - (4\sin n\frac{\pi}{2})y_{n-1} + y_{n-1} + (n-1) \qquad (3.12)$$

for $n = 3,\ldots,8$ with $y_0 = 1$

EXERCISE 3.5

Bernstein (2003) describes a model for the 10-year growth of a population of silver-back ground squirrels, and shows that the population growth is determined by

$$P_n = RP_{n-1} \qquad (3.13)$$

To construct R, field data must be used to construct a *life table* that consists of: x, the age of the cohort; l_x, the survivorship probability (the probability of surviving from 0 to x); and m_x, the fecundity (the average number of female offspring left by a female of age x). A typical life table is shown in Table 3.4.

TABLE 3.4   The Life Table for a Population of Silver-Back Ground Squirrels

x	l_x	m_x
0	1.0	0
1	0.4	0
2	0.3	2
3	0.2	3
4	0.2	3
5	0	0

Bernstein (2003) describes how this table can be used to compute the population growth rate R for discrete growth by using the following process:

1. Compute the rate of growth per generation, R_0, using

$$R_0 = \sum_{x=0}^{5} l_x m_x \qquad (3.14)$$

2. Compute the generation time, T, using

$$T = \frac{\displaystyle\sum_{x=0}^{5} l_x m_x x}{\displaystyle\sum_{x=0}^{5} l_x m_x} \qquad (3.15)$$

3. Compute the individual growth rate, r, using

$$r = \frac{\ln R_0}{T} \qquad (3.16)$$

4. Compute the population growth rate for discrete growth, R, using

$$R = e^r \qquad (3.17)$$

Using this model, simulate the population growth of a silverback ground squirrel colony assuming that the initial population is $P_0 = 100$ squirrels. The output should consist of a plot of P_n versus n for $n = 1,\ldots,10$. Be careful with precision, since these are physical data.

3.6 USING VECTOR AND MATRIX DATA

Models often involve scientific data represented in the form of vectors or matrices. Simulink provides the user with easy ways to create and use data in this form.

3.6.1 Vector and Matrix Constants

The Constant block (from the Sources library) is the means of introducing all constant data into a simulation, including vector and matrix data. When a Constant block is created, the Constant Parameters sheet gives the user an interface to enter the constant value. To enter a scalar, the number is typed into the entry block. However, if the data to be entered is a vector, the input must be entered as a list of elements, separated by blanks or commas, and enclosed within brackets (the symbols "[" and "]"). In addition, the box marked "Interpret vector parameters as 1-D" must be checked (this is the default). Figure 3.19 shows the entry of a vector constant with scalar components 1, 2, and 3, all pure numbers.

To enter a matrix constant, the "Interpret vector parameters as 1-D" box must be cleared. The data is entered as a series of rows of elements; rows are separated by semicolons, and elements are separated by blanks or commas. The entire matrix is enclosed within brackets. Figure 3.20 shows

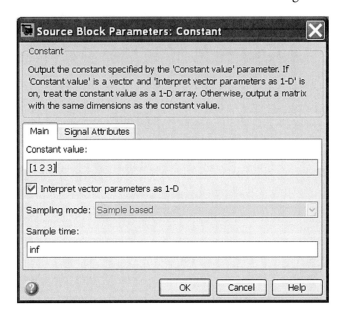

FIGURE 3.19A   Entering constant vector data.

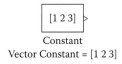

Constant
Vector Constant = [1 2 3]

FIGURE 3.19B   Entering constant vector data.

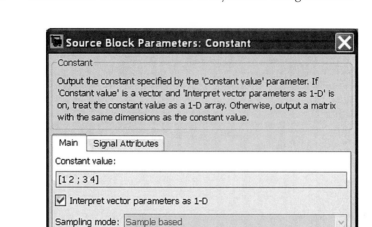

FIGURE 3.20A   Entering constant matrix data.

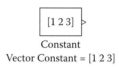

Constant

Vector Constant = [1 2 3]

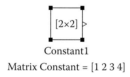

Constant1

Matrix Constant = [1 2 3 4]

FIGURE 3.20B   Icons for vector and matrix constants.

TABLE 3.5   Model Data for Model_3_5_1

i	X_i	Y_i	Z_i
1	−1.1	1.0	1
2	1.2	0.9	0.2
3	−1.0	−1.0	0.5
4	1.0	−1.1	0.1

the entry of a constant matrix, with row 1 having elements 1 and 2 and row 2 having elements 3 and 4, again, all pure numbers.

As an example, let's create a simulation for the mathematical model shown in Equation (3.18).

$$W_i^n = P_i W_i^{n-1} + 0.2 \qquad W_i^0 = 0 \qquad n = 1, \dots, 10 \qquad (3.18)$$

where

$$P_i = X_i Y_i + Z_i^2 \qquad (3.19)$$

and where the data for the model are as given in Table 3.5.

This model computes four different values for W^n, depending on which of the values of i is chosen from the table. We could create a model that has four independent layers, one for each of the four values of i. However, we can also use the vector data type to combine all four into one layer. To do this, we create a single Constant block for X_i having the vector [4.5,1.2,22.3,3.0] as its value, another for Y_i having the vector [1.0,8,–1.0,9.2], and one for Z_i having the vector [1,2.6,2.3,6]. The X_i and Y_i can be used as input to a two-input Add block and the Z_i as input to a square Math Function followed by a one-input Add block. Their outputs can be used to form the processing for the simulation, Model_3_5_1, as shown in Figure 3.21.

3.6.2 The Display Block from the Sinks Library

Examining Model_3_5_1, we see a new output block on the right side of the diagram. This is the Display block from the Sinks library whose purpose is to display its input in numerical form. This is an advantage when we want to see a number rather than a graphical point. The Display block senses its input data type and adjusts its diagram appearance to account for the data type. Thus, a scalar input will cause a Display block to display a single number, while a vector input will cause it to display a list of numbers.

For our example, the Display block automatically displays a four-number list to provide the appropriate output display for the four W_i^n values.

Example of Vector Constant Use

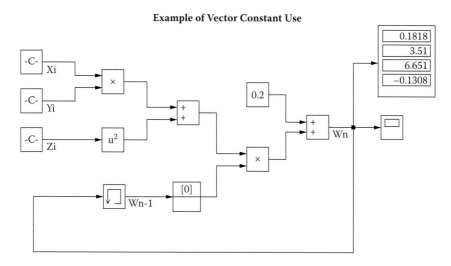

FIGURE 3.21A The simulation of the vector-constant example. In the output graph, simulation time is shown on the *x*-axis, and simulation output is shown on the *y*-axis in arbitrary units.

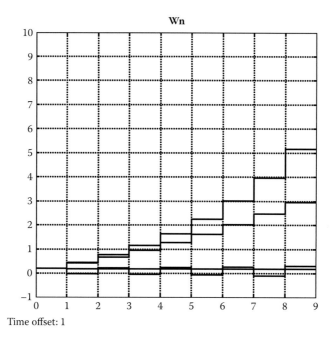

FIGURE 3.21B (See color insert following page 144.) The simulation of the vector-constant example.

3.6.3 Colors for Displaying Scope Vector Input

When we look at the Scope output of the four W_i^n inputs, we see that the Scope has correctly detected the vector input and provided us with four curves representing the four input values. What we do not see in the black-and-white image in Figure 3.21 is that the four lines are colored to enable us to discriminate among them. Just as we saw the Floating Scope use line styles to distinguish the different inputs, so does the Scope distinguish the different components of a vector input.

In Model_3_5_1, the values for $i = 1$, the leftmost of the values entered into the three vector constants, are shown in yellow, those for $i = 2$ in magenta, for $i = 3$ in cyan, and the last, $i = 4$, in red, as indicated in Table 3.6.

Suppose we change our model from Equations (3.18) and (3.19) to a different one. The modified model is shown as follows:

$$W_n = PW_{n-1} + 0.2 \quad W_0 = 0 \quad n = 1,\ldots,10 \tag{3.20}$$

where

$$P = \sum_{i=1}^{4} X_i Y_i + \sum_{i=1}^{4} Z_i^2 \tag{3.21}$$

In this model, we need to find the scalar product of the two vectors $\vec{X}$ and $\vec{Y}$ and the scalar product of $\vec{Z}$ with itself. This produces a single value for P and subsequently a single value for W_n. We can use a very similar diagram for this simulation, except that we must sum the components of the vector produced by the $X_i \times Y_i$ operation and sum those of the $Z_i \times Z_i$ operation. This can be done by inputting the vector output from these operations

TABLE 3.6   Color Assignments
for Vector Input to a Scope

Line 1	Yellow
Line 2	Magenta
Line 3	Cyan
Line 4	Red
Line 5	Green
Line 6	Dark blue
Line 7	Yellow
and so on	and so on

Second Example of Vector Constant Use

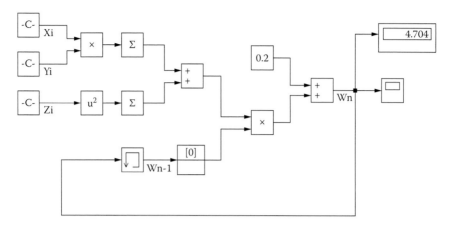

FIGURE 3.22A   Model for the second vector-constant example. In the output graph, simulation time is shown on the x-axis, and simulation output is shown on the y-axis in arbitrary units.

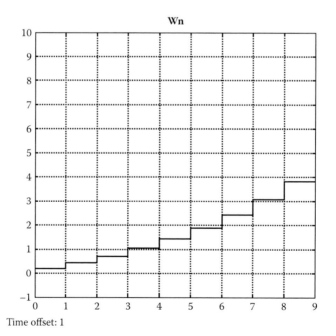

Time offset: 1

FIGURE 3.22B   Model for the second vector-constant example.

into a single-input Sum block. This block will correctly understand that we want the components summed and a single value produced on output. Model_3_6_1 in Figure 3.22 shows how we can simulate this model.

EXERCISE 3.6

Rework the Exercise 3.5 simulation of the population growth of silver-back ground squirrels to eliminate any complex networks involving multiple scalar constants, replacing them with vector constants. (Hint: We can form a vector Constant block to store the survivorship probability data, a vector Constant block to store the fecundity data, and a vector Constant block to store the cohort age data. These Constant blocks can be connected to Product block inputs, and the output will be a vector of the same data type and length as the inputs, with the products of the vector elements as its elements. A single-vector Constant block can be connected to the input of a single-input Sum block, and the output will be the sum of the vector elements. This action is exactly what we need to compute the growth rate, R, for the population growth.)

3.7 SUMMARY

In this chapter, we have seen that systems with dynamical behavior changing at regular intervals can be modeled by difference equations. This new kind of model has introduced us to the feedback diagram, a fundamental and important mechanism in modern systems. We have seen how the concept of *state* can be used in our difference equation models to retain a history of the past values of variables and parameters. When our models grow complex, we can use the concept of *abstraction* to control the complexity. We have also introduced two new data types—the vector and matrix data types—and we have seen that Simulink blocks adjust their action to the input data type.

The dynamical behavior of some systems is described by determining the values of the parameters and functions of these systems at fixed intervals in the independent variables, and the equations used are called difference equations having the form $y_n = g(y_n, y_{n-1}, y_{n-2}, \ldots, y_{n-m}, n)$. A difference equation is linear if y_n does not depend on powers of y_i or on products of y_i. The order of a difference equation is determined by how early a y_i value is needed in computing y_n.

The Memory block is used in Simulink models to provide the delayed values of its input. It stores the delayed value in an internal parameter called its state. It is used in feedback loops to provide the previous values in difference equations. The IC (Initial Condition) block is used to initialize difference equation models so that they begin with the correct initial values. An alternative approach is to use the Initial Condition parameter of the Memory block.

Layers are used to simulate higher-order difference equations, with the outputs of the layers providing later value being cascaded into the layers providing earlier values. The Subsystem block is used to abstract the details of a layer so that the top-level diagram is easy to understand and portray.

Models often involve scientific data in the form of vectors and matrices, which can be used in Simulink as values of constants and variables. Simulink determines whether the input data types are scalars, vectors, or matrices, and it adjusts block actions to suit the input data types.

REFERENCES AND ADDITIONAL READING

Bernstein, R. 2003. *Population Ecology: An Introduction to Computer Simulations.* Hoboken, NJ: John Wiley.

Dym, C. L. 2004. *Principles of Mathematical Modeling.* New York: Elsevier.

Gersting, J. 2003. *Mathematical Structures for Computer Science: A Modern Treatment of Discrete Mathematics.* New York: W. H. Freeman.

Goldberg, S. 1986. *Introduction to Difference Equations.* New York: Dover.

Haberman, R. 1977. *Mathematical Models: Mechanical Vibrations, Population Dynamics, and Traffic Flow.* Englewood Cliffs, NJ: Prentice Hall.

Huckfeldt, R., C. Kohfeld, and T. Likens. 1982. *Dynamic Modeling: An Introduction.* Beverly Hills, CA: Sage.

Keen, R. E. and J. D. Spain. 1992. *Computer Simulation in Biology: A Basic Introduction.* New York: John Wiley.

Levy, G. 2004. *Computational Finance: Numerical Methods for Pricing Financial Instruments.* New York. Elsevier.

Mooney, D., and R. Swift. 1999. *A Course in Mathematical Modeling.* Washington, DC: Mathematical Association of America.

Sedaghat, H. 2007. Difference Equations as Discrete Dynamical Equations. In *Handbook of Dynamic System Modeling.* Ed. P. A. Fishwick. Boca Raton, FL: Chapman & Hall/CRC.

CHAPTER 4

Simulation of First-Order Differential Equation Models

I N THIS CHAPTER, WE discuss systems whose behavior is described by
giving a smooth sequence of values for the relevant parameters or functions. These functions are called *continuous* functions, since the values in
any interval of the independent variable vary smoothly and the number
of values in any interval is infinite. We can define a value for a continuous
function at any point on the independent-variable axis. In this chapter,
we are going to focus on those continuous systems that have time as their
single independent variable.

Systems described by continuous functions of a single independent variable are the systems we study in introductory science courses to gain an
understanding of how theory is used in science. These systems vary from
simple physics systems through elementary chemical systems to biological
and ecological systems and beyond. They are the basic foundations of every
scientist's or technologist's career. These systems have models that are usually
described in the form of dynamical equations called *differential equations*.

4.1 WHAT IS A DIFFERENTIAL EQUATION?

Anyone who has studied calculus has already encountered differential
equations. Whenever an equation is written giving the value of a derivative, this is a differential equation.

> A differential equation is an equation containing derivatives of quantities
> of interest in the system with respect to one or more independent variables
> of the system.

4.1.1 Differential Equation Terminology

Since time is often the independent variable in dynamics, it is common to abbreviate derivatives with respect to time by the dot notation shown below. One dot over a function of time signifies a first derivative of the function with respect to time, two dots signify a second derivative with respect to time, and so on. The prime notation is often used with functions whose independent variable is a spatial dimension, such as x. A single prime sign superscripting a function of time signifies a first derivative of the function with respect to the spatial variable, and so on.

Examples of differential equations are:

$$\frac{dy(t)}{dt} = c \quad \text{or} \quad \dot{y}(t) = c \tag{4.1}$$

$$\frac{d^3 y(t)}{dt^3} + \frac{dy(t)}{dt} + a = 0 \quad \text{or} \quad \dddot{y}(t) + \dot{y}(t) + a = 0 \tag{4.2}$$

$$\frac{d^2 y(x)}{dx^2} = -[y(x)+b]\sin(x) \quad \text{or} \quad y''(x) = -[y(x)+b]\sin(x) \tag{4.3}$$

$$\left(\frac{d^2 s(t)}{dt^2}\right)^2 + \left(\frac{ds(t)}{dt}\right)^3 = 0 \quad \text{or} \quad \left(\ddot{s}(t)\right)^2 + \left(\dot{s}(t)\right)^3 = 0 \tag{4.4}$$

Differential equations that involve functions of only one independent variable are called *ordinary differential equations* (ODEs); all of the examples above are ODEs. Differential equations that involve functions of more than one independent variable are called *partial differential equations* (PDEs); Equation (4.5) is an example of a PDE. The simulation of these equations is an advanced topic, so we give only an overview of their simulation in Chapter 10.

$$\frac{\partial^2 w(x,t)}{\partial t^2} + \frac{\partial^2 w(x,t)}{\partial x^2} = xw(x,t) + t^2 \qquad (4.5)$$

The *order* of a differential equation is the order of the highest derivative present, so Equation (4.1) is first-order; Equation (4.2) is third-order; and Equations (4.3)–(4.5) are all second order. If the differential equation has powers of derivatives, the *degree* of the differential equation is the highest exponent in any term. Equations (4.1)–(4.3) and (4.5) are all of degree 1, while Equation (4.4) is of degree 3. Degree 1 equations are called *linear differential equations*, while the higher-degree equations are called *nonlinear differential equations*.

The differential equations for a system are usually formed from the laws of dynamics for the system. These involve relationships between rates of change of quantities and functions of the independent variables. From these relationships, one or more differential equations are formed that give the changes in the quantities of interest.

4.2 EXAMPLES OF SYSTEMS WITH DIFFERENTIAL EQUATION MODELS

First-order ordinary differential equations are used to represent a variety of well-known physical systems, and it is very rewarding to study them in detail. A well-known example is ordinary thermal cooling or heating. An object that is hotter (or colder) than its environment cools off (or warms up) by following physical behavior that is represented by a model consisting of a first-order differential equation.

Physicists have observed that when an object is at a temperature higher than its surroundings, the temperature of the object, $T(t)$, will change continuously to approach that of the surrounding temperature, T_E. If a small copper ball is hotter than its surroundings, it will smoothly cool off to the environmental temperature. The rate at which the temperature changes is proportional to the difference in temperatures

$$\frac{dT(t)}{dt} \propto \left| T(t) - T_E \right| \qquad (4.6)$$

and the direction of temperature change is opposite to the sign of the difference.

Thus the dynamics of the temperature of the copper ball is determined by the differential equation

$$\dot{T}(t) = -r\big(T(t) - T_E\big) \tag{4.7}$$

This is an equation with one independent variable containing a derivative raised to the first power, so we have a linear, first-order, ordinary differential equation representing the cooling copper ball system. The parameter r is a constant that represents the rate of change of the specific physical system (perhaps, a copper ball in air).

The copper ball temperature dynamics is completely determined by Equation (4.7), and this equation is so simple that a closed-form solution can be found by integration to provide an exact answer for the behavior. It is

$$T(t) - T_E = (T_0 - T_E)e^{-rt} \tag{4.8}$$

Let's work through how this is produced. The cooling dynamics is given by

$$\frac{dT(t)}{dt} = -r\big(T(t) - T_E\big) \tag{4.9}$$

which we can rewrite as

$$dT(t) = -r\big(T(t) - T_E\big)dt$$

and then as

$$\frac{1}{T(t) - T_E}dT(t) = -rdt \tag{4.10}$$

If we integrate both sides, we have

$$\int_{T_0}^{T(t)} \frac{1}{T(t) - T_E}dT(t) = \int_0^t (-r)dt \tag{4.11}$$

where T_0 is the temperature at $t = 0$. This can be reduced to

$$\ln\left(T(t)-T_{\rm E}\right)\big|_{T_0}^{T(t)}=-rt\,\big|_0^t \tag{4.12}$$

or

$$\ln\left(T(t)-T_{\rm E}\right)-\ln(T_0-T_{\rm E})=-rt \tag{4.13}$$

To eliminate the natural logarithms, we raise both sides to a power of e

$$e^{\ln(T(t)-T_{\rm E})-\ln(T_0-T_{\rm E})}=e^{-rt} \tag{4.14}$$

and rearrange by

$$\frac{e^{\ln(T(t)-T_{\rm E})}}{e^{\ln(T_0-T_{\rm E})}}=e^{-rt} \tag{4.15}$$

$$e^{\ln(T(t)-T_{\rm E})}=e^{\ln(T_0-T_{\rm E})}e^{-rt} \tag{4.16}$$

giving us the final closed-form solution

$$T(t)-T_{\rm E}=(T_0-T_{\rm E})e^{-rt} \tag{4.17}$$

We've made an effort to show all the steps in a solution by hand so that, when we run the simulation, we can compare our results with the exact solution. Of course, for this system, it would be easier to plot the exact solution rather than simulate it. But we do this so we can learn how to construct simulations in a controlled, well-understood system. When we move to a system whose solution is not well understood, we will have confidence that we are constructing a correct model.

Another simple example is a nuclear decay system (Gould and Tobochnik 1996). In a nuclear decay situation, the rate of change of nuclei from one species to another is determined by

$$\frac{dN(t)}{dt}=-\lambda N(t) \tag{4.18}$$

This is another first-order ordinary differential equation, and we can find the solution in a manner similar to the copper ball cooling system. It is

$$N(t) = N_0 e^{-\lambda t} \tag{4.19}$$

A third example comes from biology. If a population of rabbits has no predators (Australia was an example of an ecological system with no native rabbit predators), then a simple model of the dynamics of the rabbit population is the linear, first-order ODE

$$\frac{dR(t)}{dt} = g_R R(t) \tag{4.20}$$

where $R(t)$ is the rabbit population in a particular area and g_R is a population rate change for the system. Again, we find the solution to be $R(t) = R(0)e^{g_R t}$.

4.3 REWORKING FIRST-ORDER DIFFERENTIAL EQUATIONS INTO BLOCK FORM

Returning to our copper ball example, we would like to construct a simulation by starting from Equation (4.7), the initial equation describing the dynamics of the temperature change. But we don't want to have to find an analytic solution, as in our demonstration above. Instead, we want to go directly from the equation to a simulation. To do this, we manipulate Equation (4.7) in a different way than we did in finding an exact solution to this equation—one that will suit our simulation construction better. We rewrite Equation (4.7) as

$$dT(t) = -r(T(t) - T_E)dt \tag{4.21}$$

and integrate both sides

$$\int_{T_0}^{T(t)} dT(t) = \int_0^t -r(T(t) - T_E)dt \tag{4.22}$$

to arrive at

$$T(t) = T_0 - \int_0^t \left(r(T(t) - T_E) \right) dt \qquad (4.23)$$

Here is an equation for the quantity we seek, $T(t)$, in terms of the initial temperature, T_0, and the output of an integration having, as an input, the product of the thermal constant, r, along with the difference between the ball temperature and the environmental temperature. So now we can build a simulation model for this equation by using an integration to produce the second term on the right-hand side of Equation (4.23).

Before we begin, let's reverse Equation (4.23), as we did in Chapter 3, to orient the terms in the equation so that they align in a similar way to the default flow of block inputs and outputs in our Simulink® model.

$$T_0 - \int_0^t \{ r[T(t) - T_E] \} dt \rightarrow T(t) \qquad (4.24)$$

The second term on the left-hand side in Equation (4.24), the integration term, must be represented by some block, while the first term must be represented by a Constant block.

4.4 FIRST-ORDER DIFFERENTIAL EQUATION SIMULATION

How can we simulate the behavior of systems that are described by models in the form of Equation (4.24)? One way to do this is to build a Simulink simulation that uses *numerical integration methods* to provide the block we need for Equation (4.24). The numerical integration methods we need to succeed in this method are encapsulated in a Simulink block that performs the action needed to solve the integral terms in our dynamical equations—the Integrator block.

4.4.1 The Integrator Block from the Continuous Library

The fundamental simulation block for differential equations describing continuous systems is the Integrator block found in the Continuous

FIGURE 4.1   A drawing of the Integrator block's inputs and outputs.

library. It has a number of inputs and outputs, as shown in the drawing of the icon in Figure 4.1.

The Integrator block outputs the integral of its input at the current time step and simulates Equation (4.25). Its output, $y(t)$, is a function of its input, $u(t)$, and its initial condition, $y(0)$.

$$y(t) = \int_0^t u(t)dt + y(0) \qquad (4.25)$$

To perform this integration, Simulink uses one of a number of different numerical integration methods to compute the Integrator block's output, each having various advantages in particular applications. The Configuration parameters | Solver dialog box allows us to select the solver best suited to our problem.

Simulink treats the Integrator block as a dynamic system with one state, its output. The selected solver computes the output of the Integrator block at the current time step, using the current input value and the value of the state from the previous time step. The Integrator block then saves its output at the current time step for use by the solver to compute its output at the next time step.

4.4.2 Specifying Initial Values for First-Order Differential Equation Simulations

The block also provides the solver with an internal initial condition for use in computing the block's initial state at the beginning of a simulation run. The default value of the internal initial condition is 0. The block's parameter dialog box allows us to specify another value for the initial value or create an external initial value input port on the block, as seen in Figure 4.2.

First, we insert the Scope block, which will display $T(t)$, in the right-hand side of our model. Second, we add the initial temperature Constant block and the Integrator block to our model, and connect them to the

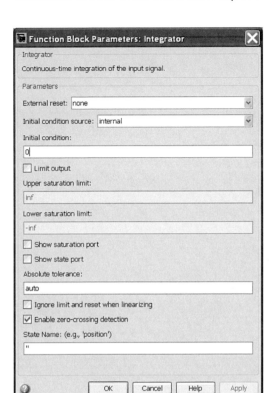

FIGURE 4.2   The Integrator block parameters.

Scope block through a Sum block. The Integrator block requires an input that is the product of the thermal constant and the difference in temperatures, so let's add a Product block to produce its input.

Now we have to supply the Product inputs. These are just the thermal constant, r, and the difference of the temperatures $T(t)$ and T_E. Let's add the thermal constant first. Since we want the thermal constant to be able to be set for each different problem we simulate, we'll leave the default value of 1.0 in the Constant block, but expect to reset it for the particular problem when we run the simulation.

Finally, we have to supply the difference input. We can do this by using a Sum block with one input set to a plus sign and the other to a minus sign. The minus part of the difference is just the room temperature, but the other part presents a problem. It's the temperature that the model is computing! So how do we supply it for input? The answer is *feedback*, as we saw in Chapter 3. We bring the output value back around into the plus input of the Sum block, as seen in Figure 4.3.

The Dynamics of Thermal Changes in an Object

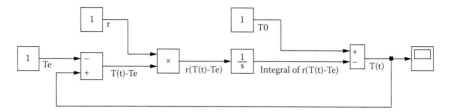

FIGURE 4.3   Model_4_1_1 of the copper ball thermal simulation.

The Dynamics of Thermal Changes in an Object

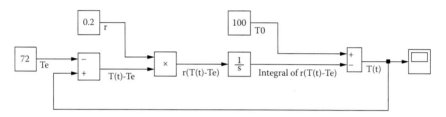

FIGURE 4.4   Model_4_1_2 of the copper ball thermal simulation.

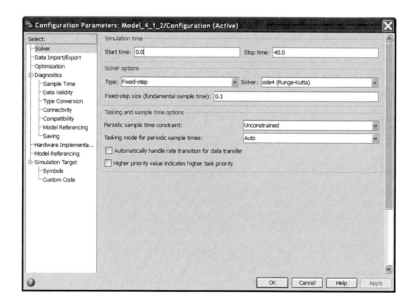

FIGURE 4.5   The configuration parameters for the copper ball simulation.

To see the simulation in action, let's choose an initial copper ball temperature of 100°F, an environment temperature of 72°F, and a thermal constant of 0.2 min^{-1}. This produces the final model shown in Figure 4.4.

We set the configuration parameters to a run of 40 min and use the default solver ode4, as seen in Figure 4.5. This produces the output shown in Figure 4.6

EXERCISE 4.1

Create a Simulink simulation of the dynamical Equation (4.26) for 0 to 10 seconds. Use an internal initial value of $Y(0) = 0$. Assume that the numbers are all precise.

$$\frac{dY(t)}{dt} = 10\cos(2\pi t) \qquad (4.26)$$

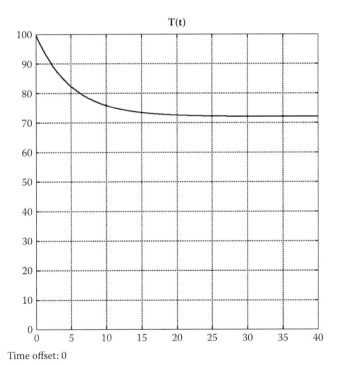

FIGURE 4.6 The temperature dynamics of the copper ball. Simulation time is shown on the x-axis in minutes, and temperature is shown on the y-axis in degrees Fahrenheit.

EXERCISE 4.2

Create a Simulink simulation of the pure mathematical Equation (4.27) for 0 to 8 seconds. Use an external initial value of $F(0) = 1$.

$$\frac{dF(t)}{dt} = t + e^{-\frac{t}{2\pi}} \cos(2\pi t) \tag{4.27}$$

EXERCISE 4.3

Keen and Spain (1992) give a classical model for the decay of the radioactive isotope ^{32}Phosphorus using the dynamical equation

$$\frac{dC(t)}{dt} = -kC(t) \tag{4.28}$$

where $C(t)$ is the concentration of the phosphorus in the medium and k is the rate constant.

1. Create a Simulink simulation that shows the decay of a sample containing an initial concentration of 500 μcuries (a curie is 3.7×10^{10} disintegrations/s and is a measure of the activity of the sample; we use it as our measure of concentration) and a rate constant $k = 0.04847$ day^{-1}. Display the output from 0 to 100 days.
2. From the simulation results, estimate the half-life of ^{32}P. The half-life is the time required for the sample activity to be reduced by 50%.

4.5 SAVING SIMULATION DATA IN MATLAB

When a simulation is constructed in Simulink, we may only need to view the results of simulation runs visually, but often we want to analyze the results using tools available only in an outside system. If the outside system is MATLAB®, we can do this by saving the output values in a form that can be used by MATLAB. This can be accomplished by using the To Workspace block from the Sinks library.

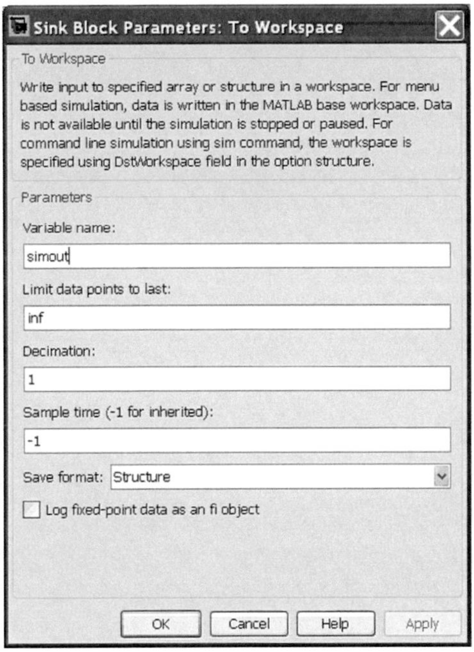

FIGURE 4.7   The To Workspace block parameters.

4.5.1 The To Workspace Block from the Sinks Library

The To Workspace block receives inputs and copies them into the workspace of the MATLAB software. Its parameters include the name of the workspace variable and the Save format, as shown in Figure 4.7.

If a variable by the name chosen does not exist in the workspace, a new variable will be created with that name. If a variable by that name already exists in the workspace, it will be overwritten by the block.

The choice for the variable's Save format is basically of two kinds: a *structure* or an *array*. A structure is a data structure consisting of a list of heterogeneous components (components whose data types can be different) that can be accessed and manipulated independently of each other. The names of the components in the list are assigned by the structure creator and are the means by which its components are fetched and manipulated. An array is a list of homogeneous components (components with identical data types) that can also be accessed and manipulated. For arrays, components are not separately named. There is an automatically created index giving the position of the component in the list that is used for fetching and manipulating.

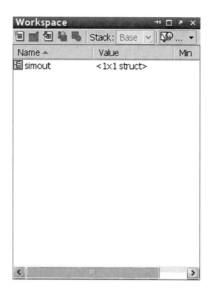

FIGURE 4.8   Workspace variable created by the To Workspace block.

Figure 4.8 shows the result of using the To Workspace block to record a sequence of 11 input values ranging from 0 to 10. There is a single variable created: the default variable named simout.

To see the components of the simout variable, we can open up the MATLAB window and double-click on the variable name in the Workspace window. An Array Editor is created with the components of simout in the Command window, and we see that the variable has three components: time, signals, and blockName. The signals component is itself a structure containing the components values, dimensions, and label. The component values is an array containing the 11 values, as can be seen in Figure 4.9. Note that these values are referenced in MATLAB through the array name simout.signals.values.

Let's use Model_4_1_2 for the copper-ball cooling problem from Figure 4.4 to record the actual temperature values in the workspace so that we can plot them in MATLAB. The data we want to transmit to MATLAB is the input sent to the Scope block, so we add a To Workspace block to our model and connect the Sum output, $T(t)$ to the To Workspace input. Let's set the To Workspace variable name to the name copperBallTemperatures and the Save format to Array, as seen in Figure 4.10.

Running the simulation and opening the MATLAB window to see the results of the Workspace change, we find our variable has been created as an array, as we see in Figure 4.11. Note that these values have not had any

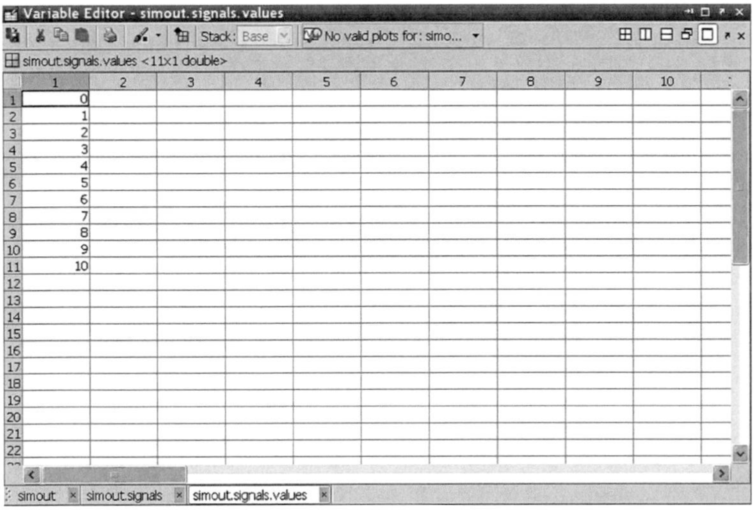

FIGURE 4.9   Variable values created by the To Workspace block.

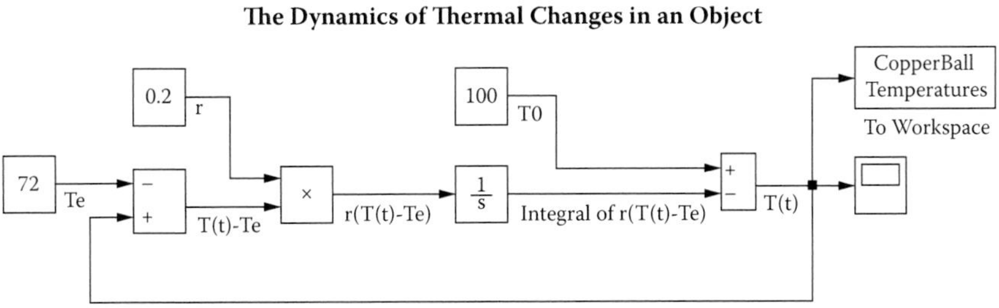

FIGURE 4.10   The temperature model for saving data to MATLAB.

precision analysis applied to them, so the simulator must be careful in their use.

Our last step is to create a plot in MATLAB (see Appendix B on plotting in MATLAB) of the temperature values in `copperBallTemperatures`, as shown in Figure 4.12.

4.5.2 Saving Simulation Data in a File

When we need to save the output variable values in a form that can be read by other programs, such as Excel, we probably need to export the data to a file in some common format. We can do this in Simulink by using the To Workspace block to transfer the output to MATLAB, and then use MATLAB commands to save the data to an ASCII file that has a standard and widely

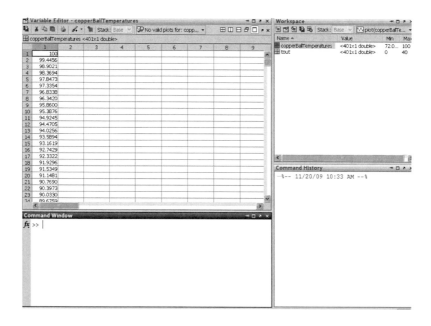

FIGURE 4.11 The values saved in the Workspace variable copperBall Temperatures.

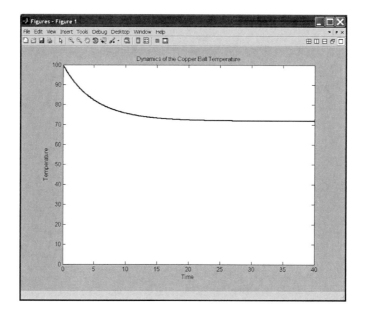

FIGURE 4.12 A plot of the copper ball temperatures using MATLAB.

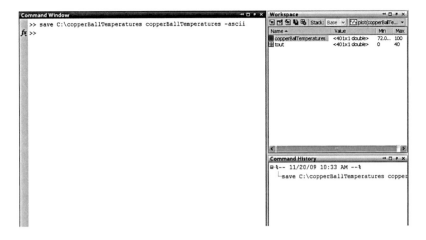

FIGURE 4.13   Using the MATLAB Save command to write an exportable file.

used format. To see a demonstration of this method, let's save the workspace variable `copperBallTemperatures` in a file in ASCII format with blank-separated values. First, we make our simulation runs and save the data to a MATLAB variable, as described previously. Then we use the command

```
save <file-path> <variable-name> -ascii
```

in the Command window to write the variable data onto a file, as shown in Figure 4.13. This command produces the file copperBallTemperatures in the C: folder whose contents are shown in Figure 4.14.

4.6 SUMMARY

In this chapter we have examined the dynamical behavior of systems described by parameters and functions that are continuous functions in the independent variable. The equations used for the important functions or parameters of these continuous systems are called differential equations. A differential equation has the form

$$\frac{d^n y(t)}{dt^n} + \frac{d^{n-1} y(t)}{dt^{n-1}} + \ldots + g(t) = 0$$

A differential equation is an ordinary differential equation if there is only one independent variable. If there is more than one, it is a partial differential equation. The degree of a differential equation is determined by the highest

FIGURE 4.14   Contents of the file `C:\copperBallTemperatures`.

power of any term in the equation. If the degree is 1, it is a linear differential equation. Otherwise, it is a nonlinear differential equation. The order of a differential equation is determined by the order of the highest derivative present. The equation given here is an nth-order differential equation.

We have seen that Simulink provides us with the Integrator block for constructing simulations of differential equations. The Integrator block of the Continuous library is used in Simulink models to simulate integrals, and its output is a function of its input and its state. The state is the current value of the integral. The state of an Integrator block can be initialized externally or internally. External initialization uses an external input port. Internal initialization uses an Integrator block parameter.

We have also seen our first use of the MATLAB software environment to expand our simulation capabilities. The To Workspace block was used to save results in the MATLAB software environment for further processing within MATLAB.

REFERENCES AND ADDITIONAL READING

Bender, E. 1978. *An Introduction to Mathematical Modeling.* New York: John Wiley.

Dym, C. 2004. *Principles of Mathematical Modeling.* New York: Elsevier Academic.

Gershenfeld, N. 1999. *The Nature of Mathematical Modeling.* Cambridge: Cambridge University Press.

Gould, H., and J. Tobochnik. 1996. *An Introduction to Computer Simulation Methods: Applications to Physical Systems.* New York: Addison Wesley.

Keen, R. E., and J. Spain. 1992. *Computer Simulation in Biology: A Basic Introduction.* New York: John Wiley.

Klee, H., 2007. *Simulation of Dynamic Systems with MATLAB and Simulink.* Boca Raton, FL: CRC Press.

Mooney, D., and R. Swift. 1999. *A Course in Mathematical Modeling.* Washington, DC: Mathematical Association of America.

Thompson, J. 2000. Understanding the United States AIDS Epidemic: A Modeler's Odyssey. In *Applied Mathematical Modeling: A Multidisciplinary Approach.* Ed. D. Shier and K. Wallenius, 41–69. Boca Raton, FL: Chapman & Hall/CRC.

CHAPTER 5

Fixed-Step Solvers and Numerical Integration Methods

W<small>E TURN NOW TO</small> the consideration of the inner workings of the solvers that we have been using in the Simulink® Memory and Integrator blocks. Throughout this chapter, we consider only solvers that use a fixed step size to solve equations. In Chapter 9, we will revisit how we can improve the efficiency of solvers by changing to a strategy of variable step size.

In this chapter, we use pure numbers throughout our discussions, so that numbers are always regarded as certain. This enables us to discuss small differences in results from different techniques without having to obscure our findings with uncertainty.

5.1 WHAT IS A SOLVER?

A solver is an implementation of a numerical algorithm that produces the output value of a block for the next time step. It tries to find the best estimate for the next output, given the information it has about the state and current inputs. As the simulation unfolds, the solver produces a stream of output values representing the action required of the block. These values are used in the rest of the simulation to form the overall behavior of the entire simulated system.

Solvers are provided in the Simulink system in the form of M-file programs. An M-file program is a program, written in the MATLAB® M-file language, that is stored on a file called an M-file and used to make a computation. The M-file language is used in MATLAB to provide a script language for system-level and user-level programming. When a simulation enters a block requiring a solver, a calculation of its output is invoked by executing the M-file program with the necessary input information.

Solvers are used by Simulink blocks that have state, so that their output is a function of their input and state. In the case of a Memory block, the output estimate is simple and precise. The output is exactly equal to the state value at each time step. And since the state value is just the input value from the preceding time step, the output value is easy to compute exactly.

$$y_n = \text{input}_{n-1} \tag{5.1}$$

For the Integrator block, computing the output of a block is basically solving the equation

$$\dot{y}(t) = f(t) \tag{5.2}$$

at each time step by estimating what the value of $y(t)$ will be at that time step. Since this is a calculation requiring an integration of the input $y(t)$ only between fixed time intervals, the output value will be

$$y_n = y_{n-1} + \int_{n-1}^{n} f_{n-1}(t)dt \tag{5.3}$$

Here, the solver must estimate the value of the integral in Equation 5.3, and this process requires a discrete-step numerical algorithm for the estimation; we must now study the algorithms that can be used for this estimation. There is no single algorithm that is best in all cases, so we consider various families of algorithms to understand how they work and when they should be used.

5.2 UNDERSTANDING THE BASICS OF NUMERICAL INTEGRATION ALGORITHMS

At the outset, it is helpful to distinguish between algorithms for definite integral evaluation and those for solving differential equations.

When a problem involves the solution of a definite integral,

$$\int_{a}^{b} f(t)dt$$

we expect an algorithm to produce a single value that is the area under the $f(t)$ curve between the points $t = a$ and $t = b$. We often have an equation form for $f(t)$, formed either from first principles of the problem or by matching an interpolant to various points in the interval $[a,b]$, and we simply want to know the area under the curve. Algorithms for definite integral evaluation are also called quadrature algorithms. If exact solutions for the integral are not known, the quadrature algorithms produce estimations of the actual value.

When a problem involves the solution to a differential equation, we generally want an algorithm that will produce a sequence of values, $f(t_0), f(t_1),...,$ that approximate the function $f(t)$, the dependent variable in a differential equation, across an interval. This sequence of values can be used to supply other functions with an approximated function, $f_{approx}(t)$.

Of these two, algorithms for the solution of differential equations are what we usually need in simulation. We want to produce the approximated function as an output signal in our block diagrams, which might be used as input to a larger system than just the differential equation solver alone. So our interest in this study is in those methods that are used to produce a sequence of values, $f_n(t)$, for the solution of a differential equation. We'll begin with the most basic and easily understood algorithm—the *Euler method.*

5.2.1 The Euler Method

Let's consider how to generate the function $y(t)$ that is the solution of the first-order ordinary differential Equation (5.4).

$$\dot{y}(t) = f(y(t),t) \tag{5.4}$$

First of all, how do we know that the solution to this first-order ODE (ordinary differential equation) exists and is unique? If it does not exist, then we are wasting our time. If it is not unique, there will be different solutions, depending on how we start the solution, and thus a simulation may not be satisfactory. So we must first inquire about these properties.

The conditions below establish these properties for us.

It is often the case that the physical systems we are simulating are so well behaved that they satisfy these conditions.

If $f'(y(t),t)$ is real, finite, single-valued, and continuous at all points $(y(t),t)$ in the region, and if $f(y(t),t)$ is real, finite, single-valued, and continuous in the region, then there is one and only one solution $y(t)$ that passes through a given point in the region.

To see how we solve Equation 5.4, suppose that we first divide the time interval over which we want the solution, $y(t)$, into a set of equal-sized steps of constant size

$$\tau = \frac{t - t_0}{n}$$

Then we can write the interval as the set $\{t_0, t_0 + \tau, t_0 + 2\tau, t_0 + 3\tau, \ldots, t_0 + n\tau\}$ and abbreviate it as $\{0,1,2,3,\ldots,n\}$. This notation gives us a way to rewrite our equations more compactly using

$$y(t = t_0 + n\tau) \equiv y_n \tag{5.5}$$

Remember that we must have a starting point in the interval to specify exactly one solution to Equation 5.4. Take that point to be one that we call an *initial value*. Then the initial value of $y(t)$ at the initial point is $y_0(t = t_0)$, or, in our compact notation, y_0. Now we can use this initial value to compute the initial value $f(y_0, t_0)$ and, ultimately, the values of $y_0, y_1, y_2, y_3, \ldots, y_n$.

We can use this notation to develop the Euler method by using a *Taylor's series expansion*.

5.2.2 Taylor's Theorem

One statement of Taylor's theorem says: If a function $y(t)$ has a Taylor's series expansion in some interval about a point t_n, then we can write the *exact* value of the function $y(t_{n+1})$ at any point t_{n+1} in the interval as

$$y(t_{n+1}) = y(t_n) + (t_{n+1} - t_n)\dot{y}(t_n) + \frac{(t_{n+1} - t_n)^2}{2!}\ddot{y}(t_n) + \frac{(t_{n+1} - t_n)^3}{3!}\dddot{y}(t_n) + \cdots$$

$$(5.6)$$

To see how Taylor's theorem works in practice, consider the case when $y(t)$ is the linear function $y(t) = 2t$, the interval is $[0,10]$, $y(t_n = 3) = 6$, and we desire the exact value at $t_{n+1} = 8$.

An application of Equation (5.6) gives

$$y(8) = 6 + (8 - 3)\cdot 2 + \frac{25}{2}\cdot 0 + \frac{125}{6}\cdot 0 + \cdots = 16 \tag{5.7}$$

since $y(t_n = 3) = 2$ and $\ddot{y}(t_n = 3) = \dddot{y}(t_n = 3) = \ldots = 0$. This is the exact value of $y(t)$ at the new point. Note that the $t_{n+1} = 8$ point is not even close to the $t_n = 3$ point, and yet we are able to extrapolate to the exact value over a long distance in one calculation step!

Another demonstration of the theorem is the function $y(t) = t^2$, an interval of $[0,6]$, $y(t_n = 2) = 4$, and we desire the exact value at $t_{n+1} = 5$. The theorem gives us

$$y(5) = 4 + (5 - 2)\cdot 4 + \frac{9}{2}\cdot 2 + \frac{27}{6}\cdot 0 + \cdots = 25 \tag{5.8}$$

since $y(t) = 2t$ and $\ddot{y}(t_n = 2) = 4, \ddot{y}(t_n = 2) = 2, \dddot{y}(t_n = 2) = \ddddot{y}(t_n = 2) = \ldots 0$. Again, the exact value is found.

Taylor's theorem is just a mathematical statement of the fact that, if we have the value of a continuous function at one point in an interval and have infinite knowledge of all the derivatives at that point, we have enough information to recover the entire function throughout the interval. For simple functions, such as our examples, these conditions are met, and we can use the derivatives to compute the infinite sum.

The problem with most physical systems that are the targets of a simulation is that we do not have all the information we need to recover the complete, exact function. So we have to resort to the use of Taylor's theorem to find *approximations* of the function.

To see how we can do this, let's rewrite Taylor's theorem in our notation.

$$y_{n+1} = y_n + \tau \dot{y}_n + \frac{\tau^2}{2!} \ddot{y}_n + \frac{\tau^3}{3!} \dddot{y}_n + \cdots \qquad (5.9)$$

Now we can develop an approximation of the value y_{n+1} by judiciously approximating terms involving $\dot{y}_n, \ddot{y}_n, \dddot{y}_n$, and higher derivatives to find a solvable equation for y_{n+1}.

The best-known approximation is to assume that the function $y(t)$ is slowly changing throughout the interval, so that all the higher-order derivatives contribute a negligible amount to the sum on the right-hand side of Equation (5.9). Then we have

$$y_{n+1} \simeq y_n + \tau \dot{y}_n \qquad (5.10)$$

This can be reduced to a familiar form by using $\dot{y}(t) = f(y(t), t)$ from Equation (5.4) to give us the expression for the Euler method.

Euler's Method for solving $\dot{y}(t) = f(y(t), t)$:
Use $y_{n+1} = y + \tau f(y(t), t)$. for computing a sequence of values for $y(t)$ at fixed-sized steps.

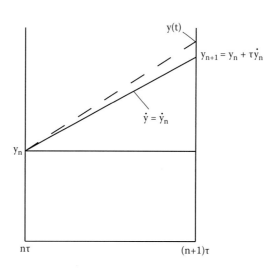

FIGURE 5.1   Graphical depiction of the Euler method.

TABLE 5.1   Steps in the Euler Method for Solving
$\dot{y}(t) = y(t), y(0) = 1, \tau = 0.2$

n	yn	$\dot{y}_n (= y_n)$	$y_{n+1} (= y_n + n \cdot \dot{y}_n)$
0.000000	1.000000	1.000000	1.200000
0.200000	1.200000	1.200000	1.440000
0.400000	1.440000	1.440000	1.728000
0.600000	1.728000	1.728000	2.073600
0.800000	2.073600	2.073600	2.488320
1.000000	2.488320		

It is executed by inserting the values for y_n and t_n into the right-hand side (RHS) and computing a new value for y_{n+1}. Iterating this procedure at each time step across the entire interval will supply the approximate values of $y(t)$ across the entire interval.

Note that Euler's method uses a known value, y_n, to predict a future value, y_{n+1}. A method of this type is called a *predictor method*. The equations used to compute unknown values from known values by such methods, like Equation (5.10), are called *open equations*.

5.2.3 A Graphical View of the Euler Method

As a different way to understand what is happening in the Euler method, we can give a geometric interpretation of the Euler method, as shown in Figure 5.1. Here we see that we are basically taking the value at the point t_n and extrapolating it to the point t_{n+1} by using a constant slope equal to the slope $\dot{y}(y_n, t_n)$. We simply extend the slope across the required interval and compute the new value by adding the change in y computed from the tangent of the angle. This emphasizes that we are predicting the new value based on data at the last known point.

If the size of the interval, τ, is very large and the solution curve is very concave or convex, this approximation is liable to be very poor. So the size of the interval must be kept small to stay within reasonable bounds. Once we have the value at the next point, the new value becomes the known value y_n, and another step can be taken. This process continues until the desired interval is spanned.

As an example of the Euler method, let's compute the solution to the equation

$$\dot{y}(t) = y(t) \tag{5.11}$$

TABLE 5.2   Exercise 5.1 Solution Template

n	Euler Method	$2e^n - n - 1$
0.000000		
0.100000		
0.200000		
0.300000		
0.400000		
0.500000		
0.600000		
0.700000		
0.800000		
0.900000		
1.000000		

for the initial condition $y(0) = 1$ by using a spreadsheet. We use a step size of 0.2 across the interval from 0 to 1, and our results are shown in Table 5.1.

EXERCISE 5.1

Using the Euler method, solve the equation $\dot{y}(t) = y(t) + t$. on the interval [0,1] manually (a spreadsheet is acceptable as a manual computation). Use $y(0) = 1$ as the initial value and $\tau = 0.1$ as the step size. Complete Table 5.2 by computing the solution at the points in the interval. Carry out the computations keeping six decimal digits. (Note that the exact solution is $2e^t - t - 1$ [Gould and Tobochnik 1996].)

Let's compare our approximate solution from Table 5.1 with the exact solution, $y(t)=e^t$, shown in Table 5.3.

The error in our Euler Method solution at
$t = 1.000000$ is $2.718282 - 2.488320 = 0.229962$.

We can graph the error between the two to get a depiction of the approximation quality, as seen in Figure 5.2.

Examination of this graph does not reassure us that the Euler method will provide us with a good solution. This brings us to a consideration of the kinds of errors that we make in solving a differential equation digitally.

TABLE 5.3 Comparison of Euler Method Solution with Exact Solution

t	y_n	$y(t) = e^t$	% error
0.0	1.000000	1.000000	0.0
0.2	1.200000	1.221403	1.8
0.4	1.440000	1.491825	3.5
0.6	1.728000	1.822119	5.2
0.8	2.073600	2.225541	6.8
1.0	2.488320	2.718282	8.5

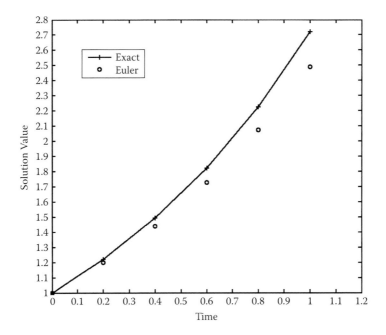

FIGURE 5.2 A graphical comparison of the Euler method errors (Euler method values shown as circles). Time and solution values are in arbitrary units.

5.3 UNDERSTANDING SOLVER ERRORS

Why is our solution above so poor? We can answer this question by considering the sources of error in the solution. The most important answer is *the choice of step size*. A large step size may produce a large error, since we are trying to predict a value too far away for the information we are using in the solver. Reducing the step size to a progressively smaller size will reduce the error we make in our predictions by shortening the distance over which we have to predict.

However, progressively reducing the step size may reveal that there are still unacceptable errors in our predicted values. This is because there are two additional sources of error in solving a differential equation by a digital computer. The first of these is *round-off error*, and the second is *local truncation error.*

Round-off error in a digital computer occurs when a number is too large or too small to be represented exactly in the numeric format used by the computer. As an example, consider the decimal fraction 0.00000000000947383826559. This number is too small to be stored by many computers, so the computer will round off the number and attempt to store it this way. The round-off process alters the number, thereby interjecting errors into any future calculation done with the number.

If we attempt to improve our solution by reducing the time step, we must make progressively smaller and smaller subdivisions of the step size to improve the accuracy of the method. But if we reduce the step size too small, then round-off error will begin to occur, and our solution will get worse, rather than better.

At a sufficiently small value of the step size, round-off typically becomes the main source of error. This occurs normally around $\tau \approx 10^{-8}$, so in very precise computations, this error must be taken into account in producing this level of precision. But this error cannot be affected by anything that a simulator can do, other than by changing computers, since it is a feature of the particular computer on which the problem is solved.

The second error source, however, *is* under the simulator's control. Local truncation error arises from the approximation made by the choice of a numerical solver method, so the simulator may be able to choose a different method to suit the problem when unacceptable errors arise.

In Equation (5.10), we truncated the expression

$$\frac{\tau^2}{2!}\ddot{y}_n + \frac{\tau^3}{3!}\dddot{y}_n + \dots$$

from our computation, thereby making an error in y_{n+1}. This is local truncation error. For the Euler method, we can see that the error we have made in this approximation is of order $O(\tau^2)$. So the error made in generating the next value will change as τ^2.

If we want to improve our approximation, we might reduce the step size. If we halve the step size and the error changes as $O(\tau^2)$, then the local truncation error is reduced by one-fourth. Thus it appears that we can make our computation more precise by reducing the step size until we get to the round-off regime, at which we cannot improve the error of the method any more.

Let's see how this works in our previous example. Let's decrease the step size to 0.1 and recompute the interval 0 to 1, with the results shown in Table 5.4.

Again, we contrast the value of the exact solution with our numerical solution in Table 5.5.

TABLE 5.4   Recomputation of Euler Method for
$\dot{y}(t) = y(t), y(0) = 1, \tau = 0.1$

n	y_n	$\dot{y}_n (= y_n)$	$y_{n+1} (= y_n + n \cdot \dot{y}_n)$
0.000000	1.000000	1.000000	1.100000
0.100000	1.100000	1.100000	1.210000
0.200000	1.210000	1.210000	1.331000
0.300000	1.331000	1.331000	1.464100
0.400000	1.464100	1.464100	1.610510
0.500000	1.610510	1.610510	1.771561
0.600000	1.771561	1.771561	1.948717
0.700000	1.948717	1.948717	2.143589
0.800000	2.143589	2.143589	2.357948
0.900000	2.357948	2.357948	2.593742
1.000000	2.593742		

TABLE 5.5   Recomparison of Euler
Method with Exact Solution for $\tau = 0.1$

t	y_n	$y(t) = e^t$	% error
0.0	1.000000	1.000000	0.0
0.1	1.100000	1.105171	0.5
0.2	1.210000	1.221403	0.9
0.3	1.331000	1.349859	1.4
0.4	1.464100	1.491825	1.9
0.5	1.610510	1.648721	2.3
0.6	1.771561	1.822119	2.8
0.7	1.948717	2.013753	3.2
0.8	2.143589	2.225541	3.7
0.9	2.357948	2.459603	4.1
1.0	2.593742	2.718282	4.6

Reducing the step size by one-half gives us an error of 2.718282 − 2.593743 = 0.124539. This is indeed a reduction in the error, but a reduction of only 0.124539/0.229962 = 0.54, not the approximately 0.25 that we expected. Why don't we get the larger reduction in error predicted for our new step size?

The reason is global truncation error. The local truncation error is the error for one step only. But the single-step errors accumulate as we move across a multistep interval. The number of steps across an interval of length T is

$$N = \frac{T}{\tau}$$

and if the error per step is $O(\tau^2)$, then the global error across the entire interval is

$$\frac{T}{\tau} O(\tau^2)$$

This is an error of $O(\tau)$, not $O(\tau^2)$ as we thought. So a reduction by half in τ only leads to a reduction in error by one-half. This is exactly what we found in our example.

5.4 IMPROVING THE BASIC ALGORITHMS

Over the years, a number of improvements to the basic numerical integration methods have been constructed. But before we look at some of these improvements, we need to discuss some notation and terminology normally used in the description of numerical methods.

The *order* of a method is an important term in describing a method. The order of a method is the exponent n of the step size in its global truncation error, $O(\tau^n)$. A first-order method has a global truncation error of $O(\tau)$, while a fourth-order method has an error $O(\tau^4)$. From this, we see that the Euler method is a first-order method.

The *number of steps* is another important descriptor of a method. The number of steps is the number of preceding y_i required to compute the next value. The Euler method is a one-step method because it only requires one preceding value, y_n, to compute the next value, y_{n+1}. As we will see later in this chapter (Section 5.4.3), there are also multistep methods in use.

Methods are also termed *explicit* or *implicit*, depending on the form of the RHS function, $f(y,t)$. If the RHS depends on the value of y that is being computed, then it is called an implicit method. If the RHS does not contain that value, it is called an explicit method. The Euler method is an explicit method, since we do not need the value of y_{n+1} on the RHS to compute y_{n+1}. We only need the value y_n. Summarizing, then, we see that the Euler method is a first-order, one-step, explicit method.

There are many ways in which the basic algorithms can be improved. We could take additional terms in the Taylor's expansion into account to make the estimate more precise, or take into account the values at earlier time steps as well, or correct our errors as we move across the solution interval. All of these have been the subject of work by mathematicians and scientists to improve the precision of solvers.

5.4.1 *Runge–Kutta* Methods

The best known of the improved methods are the *Runge–Kutta* methods, which are described below. This family of methods has been developed in a sufficiently general way that many earlier methods can be shown to be instances of this family. The derivation of these methods is accomplished by extensive analysis using the same kind of Taylor's series expansion technique that we saw in the Euler method. Since our main goal in this book is to see how these solvers are used in simulation, we will not examine the details of the development of the *Runge–Kutta* methods. Instead, we will list the final equations for y_{n+1} and briefly describe their properties and usage. Where the *Runge–Kutta* method being discussed is known by another name, we point this out.

First-Order Runge–Kutta Methods

The first-order *Runge–Kutta* method has the equation

$$y_{n+1} = y_n + \tau f(y_n, n) \tag{5.12}$$

and is identical to the Euler method.

Second-Order Runge–Kutta Methods

One representation of the second-order RK method has the equation

$$y_{n+1} = y_n + \tau f(y_{n+\frac{1}{2}}, n + \tfrac{1}{2}) \tag{5.13}$$

and is identical to earlier methods called the Euler midpoint method or the modified Euler method. The value for

$$y_{n+\frac{1}{2}}$$

is computed using the standard Euler method and a half-size step.

$$y_{n+\frac{1}{2}} = y_n + \frac{1}{2}\tau f(y_n, n) \tag{5.14}$$

Then the

$$y_{n+\frac{1}{2}}$$

value is used in the RHS of the standard Euler method for y_{n+1} (Equation 5.13).

Third-Order Runge–Kutta Methods

One way of writing the third-order RK method has the equation

$$y_{n+1} = y_n + \tau \frac{1}{4}(k_1 + 3k_3) \tag{5.15}$$

where

$$k_1 = f(y_n, t_n) \tag{5.16}$$

$$k_2 = f(y_{n+\frac{1}{3}}, t_n), \text{ with } y_{n+\frac{1}{3}} = y_n + \frac{1}{3}\tau k_1 \tag{5.17}$$

$$k_3 = f(y_{n+\frac{2}{3}}, t_n), \text{ with } y_{n+\frac{2}{3}} = y_n + \frac{2}{3}\tau k_2 \tag{5.18}$$

This method is identical to Heun's method. It amounts to the Euler method that uses a slope equal to the weighted average of the known slope at t_n and the predicted slope at

$$t_{n+\frac{2}{3}}$$

with proportions of 1:3. The slope at

$$t_{n+\frac{2}{3}}$$

is predicted by using an Euler method to get the slope at

$$t_{n+\frac{1}{3}}$$

and then the slope at

$$t_{n+\frac{1}{3}}$$

is reused in another Euler method to get the slope at

$$t_{n+\frac{2}{3}}$$

Note that the divisor outside the parentheses of Equation (5.15) must be equal to the sum of the weights inside the parentheses.

Fourth-Order Runge–Kutta Methods

The most well-known and often-used form of the fourth-order RK methods has the equation

$$y_{n+1} = y_n + \tau \frac{1}{6}(k_1 + 2k_2 + 2k_3 + k_4) \qquad (5.19)$$

where

$$k_1 = f(y_n, t_n) \qquad (5.20)$$

$$k_2 = f(y_{n+\frac{1}{2}}, t_n), \ \text{with} \ y_{n+\frac{1}{2}} = y_n + \frac{1}{2}\tau k_1 \qquad (5.21)$$

$$k_3 = f(y_{n+\frac{1}{2}}, t_n), \text{ with } y_{n+\frac{1}{2}} = y_n + \frac{1}{2}\tau k_2 \qquad (5.22)$$

$$k_4 = f(y_{n+1}^{\text{pred}}, t_n), \text{ with } y_{n+1}^{\text{pred}} = y_n + \tau k_3 \qquad (5.23)$$

This method is the workhorse of simulation and is probably used more often than any other. Note that it is a weighted average of four slopes: the known slope at $t_n - k_1$, two estimated slopes at $t_n - k_2$ and k_3, and a predicted slope at $t_{n+1} - k_4$.

EXERCISE 5.2

Using the fourth-order *Runge–Kutta* method, solve equation $\dot{y}(t) = y(t) + t$ on the interval $[0,1]$ manually, just as in Exercise 5.1. Use the same initial value and step size used in Exercise 5.1. Construct Table 5.6 by computing the solution at the points in the interval. Carry out the computations keeping six decimal digits (Gould and Tobochnik 1996).

5.4.2 Corrector Methods

Recall that the Euler method is a predictor method. It uses current data (the values of y_n and $\dot{y}_n$) to predict where the function y_{n+1} will be at the end of the next interval. Suppose we consider whether there is a method for correcting our prediction. If there is, we could apply the correction to

TABLE 5.6   Exercise 5.2 Solution Template

n	4th-Order RK Method	$2e^n - n - 1$	error
0.000000			
0.100000			
0.200000			
0.300000			
0.400000			
0.500000			
0.600000			
0.700000			
0.800000			
0.900000			
1.000000			

improve the approximation that the Euler method makes. There are such correction methods, and the one we examine is called the *Euler predictor-corrector* (EPC) method. Geometrically, we can understand the method by examining Figure 5.3.

In the Euler method, the prediction from the values of y_n and $\dot{y}_n$ causes us to miss the exact point. If we compute the slope at the *predicted* point, it has a different value from that of the slope at the previous point. We show below that the *average* of the two slopes produces a line very close to the chord in the diagram. The slope of the chord produces a much better new value than either slope alone would.

Suppose we write the Taylor's series expansion of the $y(t)$ function at the points t_n and t_{n+1}. We first write an equation for $y(t_{n+1})$ expanding about the point t_n. This is

$$y(t_{n+1}) = y(t_n) + \tau \dot{y}(t_n) + \frac{\tau^2}{2!}\ddot{y}(t_n) + \frac{\tau^3}{3!}\dddot{y}(t_n) + \cdots \qquad (5.24)$$

Next we write an equation for $y(t_n)$ expanding about the point t_{n+1}.

$$y(t_n) = y(t_{n+1}) - \tau \dot{y}(t_{n+1}) + \frac{\tau^2}{2!}\ddot{y}(t_{n+1}) - \frac{\tau^3}{3!}\dddot{y}(t_{n+1}) + \cdots \qquad (5.25)$$

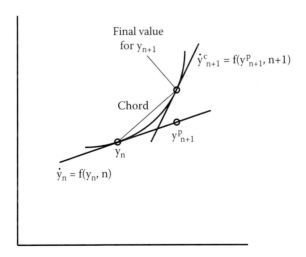

FIGURE 5.3 Graphical portrayal of the EPC method.

Now let's compute the slope of the chord connecting these two points. It is

$$\frac{y(t_{n+1}) - y(t_n)}{\tau} \tag{5.26}$$

Substituting Equations (5.24) and (5.25) into (5.26), we have

$$\frac{y(t_{n+1}) - y(t_n)}{\tau} = \frac{1}{\tau} \left[y(t_n) + \tau \dot{y}(t_n) + \frac{\tau^2}{2!} \ddot{y}(t_n) + \frac{\tau^3}{3!} \dddot{y}(t_n) + \cdots \right]$$

$$- \left[y(t_{n+1}) - \tau \dot{y}(t_{n+1}) + \frac{\tau^2}{2!} \ddot{y}(t_{n+1}) - \frac{\tau^3}{3!} \dddot{y}(t_{n+1}) + \cdots \right] \tag{5.27}$$

Combining the $y(t_n)$ and $y(t_{n+1})$ terms on the RHS and bringing them to the LHS gives

$$2 \frac{y(t_{n+1}) - y(t_n)}{\tau} =$$

$$\left[\dot{y}(t_n) + \frac{\tau}{2!} \ddot{y}(t_n) + \frac{\tau^2}{3!} \dddot{y}(t_n) + \cdots \right] + \left[\dot{y}(t_{n+1}) - \frac{\tau}{2!} \ddot{y}(t_{n+1}) + \frac{\tau^2}{3!} \dddot{y}(t_{n+1}) + \cdots \right] \tag{5.28}$$

Rearranging terms on the RHS produces

$$2 \frac{y(t_{n+1}) - y(t_n)}{\tau} =$$

$$\left[\dot{y}(t_n) + \dot{y}(t_{n+1}) + \frac{\tau}{2!} \ddot{y}(t_n) - \frac{\tau}{2!} \ddot{y}(t_{n+1}) + \frac{\tau^2}{3!} \dddot{y}(t_n) + \frac{\tau^2}{3!} \dddot{y}(t_{n+1}) + \cdots \right] \tag{5.29}$$

Truncating all terms in $\ddot{y}(t_n), \dddot{y}(t_n)$, and higher derivatives, we have

$$\text{chord slope} = \frac{1}{2}\left(\dot{y}(t_n) + \dot{y}(t_{n+1})\right) \tag{5.30}$$

which shows that the slope of the chord is the average of the slopes at t_n and t_{n+1}.

To make use of this improved slope, we need a way to compute $\dot{y}(t_{n+1})$. This value can be computed by using the previously described Euler predictor method to compute y_{n+1} and substitute it into $f(y_{n+1}, n+1)$ to find $\dot{y}(t_{n+1})$. Once we have the predicted value of $\dot{y}(t_{n+1})$, we can use it together with $\dot{y}(t_n)$ to compute the chord slope. Finally, we use the chord slope in the standard Euler method, replacing $\dot{y}(t_n)$ by the chord slope, to compute the final y_{n+1}.

So the Euler predictor-corrector method algorithm proceeds as follows:

1. Compute $y_{n+1}^{predicted}$ from y_n and $\dot{y}_n$ using the standard Euler method.

2. Compute $\dot{y}(t_{n+1})$ from $f(y_{n+1}^{predicted}, n+1)$.

3. Compute $\dot{y}_n^{EPC}$, the average of the slopes $\dot{y}_n$ and $\dot{y}_{n+1}$.

4. Recompute y_{n+1} from y_n and $\dot{y}_n^{EPC}$ using the standard Euler method.

To demonstrate the EPC method, let's compute Table 5.7 of the solution to the equation $yY(t) = y(t)$ with the initial condition $y(0) = 1$ and a step size of $\tau = 0.1$. Again, we can compare the solution to the exact values with results in Table 5.8.

We have an error of $2.718282 - 2.714081 = 0.004201$. This is an order-of-magnitude improvement over the Euler method, justifying the additional computational work in computing the EPC values. Figure 5.4 shows a graphical presentation of the predictor and predictor-corrector solutions.

The EPC method is a second-order method, and it requires that we evaluate the RHS at y_{n+1} to compute $\dot{y}_{n+1}$ so that we can compute the average of the two slopes, $\dot{y}(t_n)$ and $\dot{y}(t_{n+1})$, to find the chord slope. So the value of y_{n+1} is required during the computation of y_{n+1}. This makes the EPC method an implicit method. The EPC method is therefore a second-order, one-step, implicit method.

TABLE 5.7   Computation of EPC Method Solution

n	yn	$\dot{y}_n(=y_n)$	y_{n+1}^{EPC}	$\dot{y}_{n+1}=y_{n+1}$	$\dot{y}_n^{EPC}$	Final y_{n+1}
0.000000	1.000000	1.000000	1.100000	1.100000	1.050000	1.105000
0.100000	1.105000	1.105000	1.215500	1.215500	1.160250	1.221025
0.200000	1.221025	1.221025	1.343128	1.343128	1.282076	1.349233
0.300000	1.349233	1.349233	1.484156	1.484156	1.416694	1.490902
0.400000	1.490902	1.490902	1.639992	1.639992	1.565447	1.647447
0.500000	1.647447	1.647447	1.812191	1.812191	1.729819	1.820429
0.600000	1.820429	1.820429	2.002472	2.002472	1.911450	2.011574
0.700000	2.011574	2.011574	2.212731	2.212731	2.112152	2.222789
0.800000	2.222789	2.222789	2.445068	2.445068	2.333928	2.456182
0.900000	2.456182	2.456182	2.701800	2.701800	2.578991	2.714081
1.000000	2.714081					

TABLE 5.8   Comparison of EPC Method Solution with Exact Solution

t	y_n	$y(t)=e^t$	% error
0.0	1.000000	1.000000	0.0
0.1	1.105000	1.105171	0.0
0.2	1.221025	1.221403	0.0
0.3	1.349233	1.349859	0.0
0.4	1.490902	1.491825	0.1
0.5	1.647447	1.648721	0.1
0.6	1.820429	1.822119	0.1
0.7	2.011574	2.013753	0.1
0.8	2.222789	2.225541	0.1
0.9	2.456182	2.459603	0.1
1.0	2.714081	2.718282	0.2

EXERCISE 5.3

Using the Euler predictor-corrector method, solve equation $\dot{y}(t)=y(t)+t$ on the interval [0,1] manually, just as in Exercise 5.1. Use the same initial value and step size as that used in Exercise 5.1. Construct Table 5.9 by computing the solution at the points in the interval. Carry out the computations keeping six decimal digits (Gould and Tobochnik 1996).

5.4.3 Multistep Methods

We mentioned earlier that some methods were multistep methods. Examples can be found in methods that use interpolation to solve

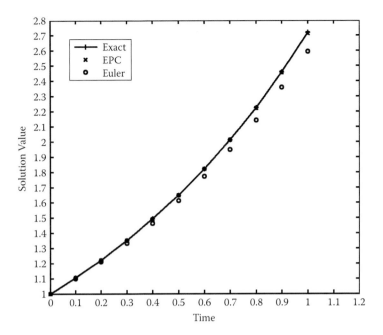

FIGURE 5.4   A graphical comparison of the EPC and Euler methods with the exact solution to $\dot{y}(t) = y(t)$, $y(0) = 1$, $\tau = 0.1$.

differential equations. Interpolation methods are common in definite integral evaluation when a few points of the curve are known, and an interpolant function is used in the integral. Solver methods can also make use of curve fitting, if the method has previous values available to it.

As an example of a simple multistep method, assume that we have available to us the points y_{n-1} and y_n. Then we can use a *Newton-Gregory interpolant* on the points y_{n-1}, y_n, and y_{n+1} to construct the equation

$$y_{n+1} = y_{n-1} + \tau \frac{1}{3}(f(y_{n+1}, t_{n+1}) + 4f(y_n, t_n) + f(y_{n-1}, t_{n-1}) \qquad (5.31)$$

This is an implicit, two-step method for computing the next point. It is equivalent to the *Newton-Coates* quadrature formula used in evaluating definite integrals. Note that multistep methods are not self-starting. The initial value y_0 is not sufficient to generate y_1, so the startup of a two-step method requires that a non-multistep method, such as a *Runge–Kutta* method, be used to generate the minimum number of values to start the method.

TABLE 5.9   Exercise 5.3 Solution Template

n	$2e^n - n - 1$	Euler	error	EPC	error
0.000000					
0.100000					
0.200000					
0.300000					
0.400000					
0.500000					
0.600000					
0.700000					
0.800000					
0.900000					
1.000000					

Another example of a two-step method is the *Adams–Moulton* method, defined by the equation

$$y_{n+1} = y_n + \tau \frac{1}{12}\left(5f(y_{n+1},t_{n+1}) + 8f(y_n,t_n) - f(y_{n-1},t_{n-1})\right) \qquad (5.32)$$

This is one of a family of methods and is more commonly represented by the fourth-order, four-step Adams–Moulton method.

The four-step Adams–Moulton method works by fitting a cubic equation to four slope values—$f(y_{n-3},t_{n-3})$, $f(y_{n-2},t_{n-2})$, $f(y_{n-1},t_{n-1})$, $f(y_n,t_n)$—by using the *Lagrange polynomial* method as a predictor stage to yield the equation

$$y_{n+1}^{\text{pred}} = y_n + \tau \frac{1}{24}\left(-9f(y_{n-3},t_{n-3}) + 37f(y_{n-2},t_{n-2}) - 59f(y_{n-1},t_{n-1}) + 55f(y_n,t_n)\right) \qquad (5.33)$$

This is followed by a corrector stage in which a cubic equation is fitted to the slope values— $f(y_{n-2},t_{n-2})$, $f(y_{n-1},t_{n-1})$, $f(y_n,t_n)$, $f(y_{n+1}^{\text{predicted}},t_{n+1})$ —using a Lagrange polynomial method again to yield the equation

$$y_{n+1} = y_n + \tau \frac{1}{24}\left(f(y_{n-2},t_{n-2}) - 5f(y_{n-1},t_{n-1}) + 19f(y_n,t_n) + 9f(y_{n+1}^{\text{pred}},t_{n+1})\right) \qquad (5.34)$$

5.5 FIXED-STEP SOLVERS IN THE SIMULINK SOFTWARE

Simulink supplies a number of continuous solvers that implement the most important fixed-step methods we have just seen. We list the solvers and summarize the methods they implement. More details can be found in the Simulink documentation.

ode1

> This solver implements the standard, first-order Euler method. To see how this is done, let's examine the Simulink M-file code for the solver. For our purposes, the relevant part of this code is boldfaced. Even without experience in the script language, we can see that the variable Y(:,i+1) is being computed by adding the increment h(i)*feval(odefun,tspan(i),Y(:,i),v arargin{:}) to the variable Y(:,i), which is the $y_n + \tau f(y_n, t_n)$ term from our discussion of the Euler method. The function feval applies the function named odefun to the remaining arguments and the loop increments across the desired interval. See Figure 5.5 for an example of MATLAB code implementing the ODE solver.

ode2

> This solver implements the second-order EPC method.

ode3

> This solver implements the *Bogacki–Shampine* variant of the *Runge–Kutta* method.

ode4

> This solver implements the classical fourth-order *Runge–Kutta* method.

ode5

> This solver implements the *Dormand–Prince* method, which is an implicit, fixed-step, continuous solver.

```
function Y = ode1(odefun,tspan,y0,varargin)
%ODE1 Solve differential equations with a non-adaptive method of order 1.
%  Y = ODE1(ODEFUN,TSPAN,Y0) with TSPAN = [T1, T2, T3, ... TN] integrates
%  the system of differential equations y' = f(t,y) by stepping from T0 to
%  T1 to TN. Function ODEFUN(T,Y) must return f(t,y) in a column vector.
%  The vector Y0 is the initial conditions at T0. Each row in the solution
%  array Y corresponds to a time specified in TSPAN.
...
neq = length(y0);
N = length(tspan);
Y = zeros(neq,N);
Y(:,1) = y0;
for i = 1:N-1
Y(:,i+1) = Y(:,i) + h(i)*feval(odefun,tspan(i),Y(:,i),varargin{:});
end
...
```

FIGURE 5.5   MATLAB code implementing the ODE solver.

ode14x

This solver implements Newton's method and extrapolation.

EXERCISE 5.4

1. Using the Euler method, choose an appropriate solver and simulate the equation $\dot{y}(t)=1+2\pi\cos 2\pi t$ on the interval $[0,1.8]$
   with $y(0) = 0$ as the initial value. Use a fixed step size $\tau = 0.1$
   first and then repeat the simulation for a step size $\tau = 0.1$. Give
   the value for $y(1.8)$ to six decimal digits at each step size.
2. Choose an appropriate solver that uses the fourth-order *Runge–
   Kutta* method and repeat the simulation above. Compare the
   results of the different solvers and step sizes.

5.6 SUMMARY

We investigated the nature of basic solvers in this chapter and found that a variety of algorithmic methods are implemented, leading to different solvers available to Simulink users. A solver is an implementation of a numerical algorithm that produces an output value of a block for the next time step. Solvers are provided in the Simulink system in the form of M-file programs for blocks that require them for correct operation. Solvers are generally used for Simulink blocks that have state, so that their output is a function of their input and state. Solvers can be fixed-step or variable-step solvers.

Euler's method for solving $\dot{y}(t) = f(y(t),t)$ is the application of the equation $y_{n+1} = y_n + \tau f(y(t),t)$ at each time step to estimate the value of $y(t)$ at the next time step. It is a predictor method. It uses open equations to compute unknown values from known values. The Euler method is a first-order, one-step, explicit method.

We also examined the *Runge–Kutta* solver and methods for improvement in the basic solvers. The *Runge–Kutta* methods are a family of methods. The first-order RK method is identical to the Euler method, the second-order RK method is identical to the Euler midpoint method, and the third-order RK method is identical to Heun's method. The fourth-order RK method is the best known and most often used RK method. It uses the equation

$$y_{n+1} = y_n + \tau \frac{1}{6}(k_1 + 2k_2 + 2k_3 + k_4)$$

to estimate the next value.

We discussed the concept of error arising because of our algorithmic truncations of higher-order terms. The kinds of errors that are encountered in simulation are round-off and local truncation errors. Global truncation error is the accumulated local truncation error across the interval of simulation. Error properties of numerical algorithms are indicated by giving the big-O function of the step-size power, $O(\tau^n)$.

Corrector methods apply a correction to improve the approximation that a predictor makes. The Euler predictor-corrector method is a second-order, one-step, implicit corrector method. Multistep methods use multiple values from prior time steps to estimate the next value. They often use interpolation methods to form an equation for extrapolation. The fourth-order, four-step Adams–Moulton method fits a cubic equation to four time-step values to estimate the next value.

Simulink supplies a number of fixed-step solvers that implement the most important numerical methods (ode1, ode2, ode3, ode4, ode5, and ode14x). These are chosen by setting the Solver parameter value on the Configuration Parameters page of a simulation.

REFERENCES AND ADDITIONAL READING

Chapra, S. C., and R. Canale. 2002. *Numerical Methods for Engineers: With Software and Programming Applications*. New York: McGraw-Hill.

Esquembre, F., and W. Christian, Ordinary Differential Equations. In *Handbook of Dynamic System Modeling*. Ed. P. Fishwick. Boca Raton, FL: Chapman & Hall/CRC.

Garcia, A. 2000. *Numerical Methods for Physics*. Upper Saddle River, NJ: Prentice Hall.

Gershenfeld, N. 1999. *The Nature of Mathematical Modeling*. Cambridge: Cambridge University Press.

Gould, H., and J. Tobochnik. 1996. *An Introduction to Computer Simulation Methods: Applications to Physical Systems*. New York: Addison Wesley.

Higham D. J., and N. Higham. 2005. *MATLAB Guide*. Philadelphia: Society for Industrial and Applied Mathematics.

Klee, H. 2007. *Simulation of Dynamic Systems with MATLAB and Simulink*. Boca Raton, FL: CRC Press.

Moler, C. B. 2004. *Numerical Computing with MATLAB*. Philadelphia: Society for Industrial and Applied Mathematics.

Mooney, D., and R. Swift. 1999. *A Course in Mathematical Modeling*. Washington, DC: Mathematical Association of America.

The MathWorks, Inc. 2007. http://www.mathworks.com.

CHAPTER 6

Simulation of First-Order Equation Systems

IN THIS CHAPTER, WE examine systems whose dynamics are described by giving a *system* of first-order equations. Physical systems whose behavior consists of a number of interacting processes will often have their dynamics described by a set of first-order equations, with one equation describing each individual process. Usually, the effect of these processes on each other is provided by including interaction terms in the equations. This gives rise to a system of equations that must be simulated as a whole, since the interactions are simultaneous with the dynamics of the individual processes. As we will see in Chapter 9, higher-order differential equations can be transformed into systems of first-order equations, thereby producing another source of coupled first-order equations.

Sometimes, interactions are small enough that, in a first approximation, the interaction terms can be dropped and the dynamics solved for each process independently of the others, thus making it possible to find analytic solutions. But simulation can be used even when such decoupling is not possible and analytic solutions are not available.

Systems of equations are often represented in matrix form so that matrix methods can be used in their solution. In particular, if the equations of a system are linear with constant coefficients and a forcing function, the system can be recast into *state variables*, and the resulting set of linear equations written as a set of matrix equations in terms of special matrices representing system inputs and outputs. These methods, called *state space methods*, are commonly studied in engineering.

Continuing our previous approach of analyzing difference equations separately from differential equations, we begin by considering systems described by a set of first-order difference equations.

6.1 WHAT IS A FIRST-ORDER DIFFERENCE EQUATION SYSTEM?

A system of first-order difference equations is just a set of first-order difference equations where any equation of the set may include the outputs from the other equations. These terms are called *interaction* terms and the equations are called *coupled* equations. Interaction terms are mechanisms by which the dynamics of a process can be affected by the dynamics of the other processes.

A system of first-order difference equations is a set of n first-order difference equations describing the dynamics of the functions $y_1^k, y_2^k, \ldots, y_n^k$ in the form

$$y_1^k = f_1\left(y_1^{k-1}, y_2^{k-1}, \ldots, y_n^{k-1}, k\right) \tag{6.1}$$

$$y_2^k = f_2\left(y_1^{k-1}, y_2^{k-1}, \ldots, y_n^{k-1}, k\right) \tag{6.2}$$

$$\ldots$$

$$y_n^k = f_n\left(y_1^{k-1}, y_2^{k-1}, \ldots, y_n^{k-1}, k\right) \tag{6.3}$$

6.2 EXAMPLES OF FIRST-ORDER DIFFERENCE EQUATION SYSTEMS

One example is a survey process organized as successive interviews of a set of panel members on their opinion on a particular question (Goldberg 1986). The survey question is a yes or no question such as, "Do you believe the economy is improving?" We can formulate a model for this process by considering that, at the nth interview, p_n is the probability that a responder will answer "yes" and q_n is the probability that a responder will answer "no." If we denote the probability that a "yes" responder will change to a "no" responder at the next interview by α and that a "no" responder will change to a "yes" responder by β, then Equations (6.4) and (6.5) constitute a system of first-order difference equations describing the dynamics of the panelists' opinions.

$$p_k = (1-\alpha)p_{k-1} + \beta q_{k-1} \qquad (6.4)$$

$$q_k = (1-\beta)q_{k-1} + \alpha p_{k-1} \qquad (6.5)$$

Examining these equations, we see that each difference equation has an interaction term on the right-hand side (RHS) by which the responders of the opposite opinion can change the dynamics of the probability.

6.3 SIMULATING A FIRST-ORDER DIFFERENCE EQUATION SYSTEM

We can create a simulation of a system of first-order difference equations by proceeding as we did in Chapter 3. In this case, we will have two *layers* in our simulation model—one for the p_k equation and one for the q_k equation. But we would like to display these outputs in the same scope, so let's insert a single scope. Implementing each of the equations up to the second right-hand-side term and a single scope produces Model_6_1_1, shown in Figure 6.1.

Now we have to add the missing second term to each layer. What we must do here is to feed the q_{k-1} signal back into the p_k layer and add it into the input of the Add block. But we must be careful, since the signal used in computing p_k is not the output of the q_k layer—it is the output from the previous time step. This means we must use the output of the q_{k-1} Memory block to get the last time-step output. We must do the same for the p_{k-1} signal and the q_k layer, and, when we are done, we have a model with two *cross-coupled* layers, as shown in Figure 6.2.

6.3.1 The Two-Input Scope Block

Next, we modify the scope so that it has two inputs. Double-left-clicking the Scope icon causes the Scope output window to appear. On the toolbar of this window, there is a Parameters menu icon, as shown in Figure 6.3. Selecting the Parameters icon gives us the Scope parameters window shown in Figure 6.4, in which we can set the number of axes to 2, which creates two inputs automatically. The Scope output window changes to the one shown in Figure 6.5.

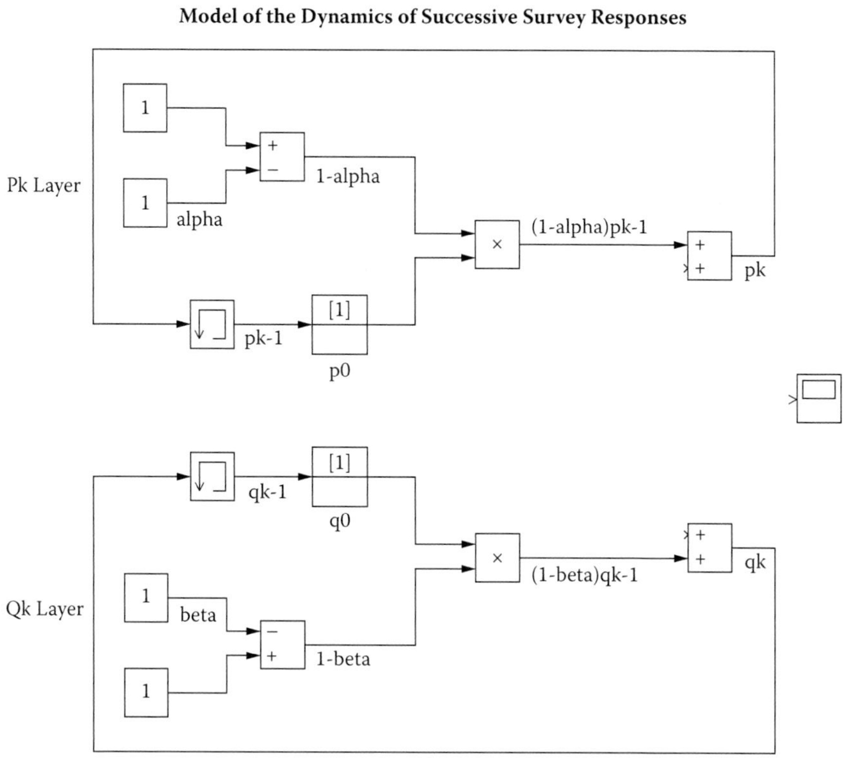

Model of the Dynamics of Successive Survey Responses

FIGURE 6.1   Model_6_1_1 for the panel survey opinion dynamics model.

We can now connect the outputs p_k and q_k to the inputs of the Scope and display them in the same window. Before doing this, let's make two other changes to produce a better output. First, let's insert two IC (Initial Condition) blocks at the Scope inputs that will have the values of p_0 and q_0, so that our output will display the initial values of p_k and q_k at time 0. Then, let's set the axes range to 0 and 1, since these outputs are both probabilities and must be in this range. This produces Model_6_1_3, shown in Figure 6.6.

Considering the case when the panel members are divided at the start ($p_0 = 0.3$ and $q_0 = 0.7$), with $\alpha = 0.2$ and $\beta = 0.3$, and then running the simulation using the Fixed-Step Discrete solver with a time step of 1 for 0–10 interviews gives the output seen in Figure 6.7.

6.3.2 Algebraic Loops

An important issue to consider now that we are constructing crosscoupled-layer models is the creation of *algebraic loops*. This is the term given in

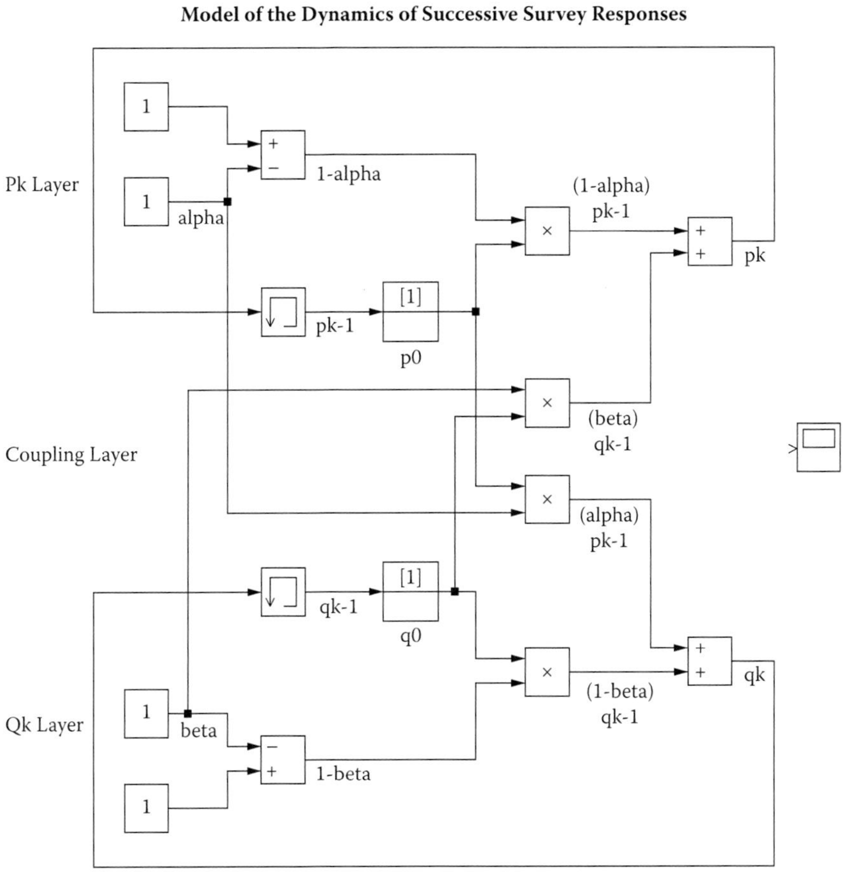

Model of the Dynamics of Successive Survey Responses

FIGURE 6.2   Model_6_1_2 for the opinion dynamics model with crosscoupled layers.

Simulink® for situations where the value propagation algorithm fails due to circular paths in the simulation diagram. To illustrate the problem, consider the system modeled by Equations (6.6) and (6.7).

$$x_k = 2y_k - 10 \tag{6.6}$$

$$y_k = 4 + 2x_k \tag{6.7}$$

A model for these equations is shown in Figure 6.8.

If we run this model with a Fixed-Step Discrete solver and a time step of 1, Simulink will stop and print out a message in the model window informing the simulator that an algebraic loop has been detected and

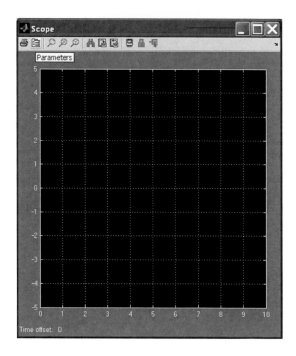

FIGURE 6.3   Scope block parameters location.

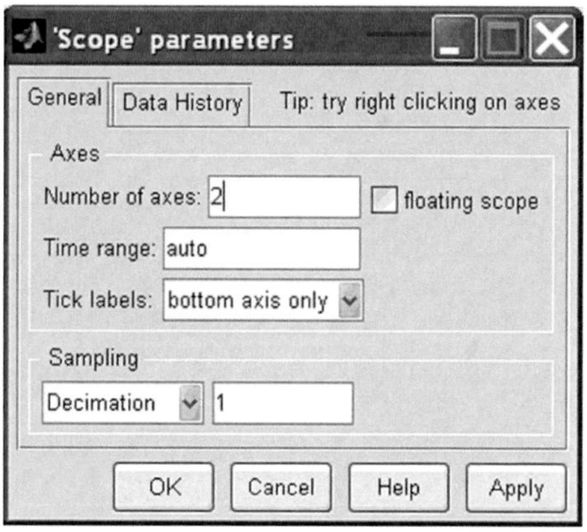

FIGURE 6.4   Creating two inputs for a scope.

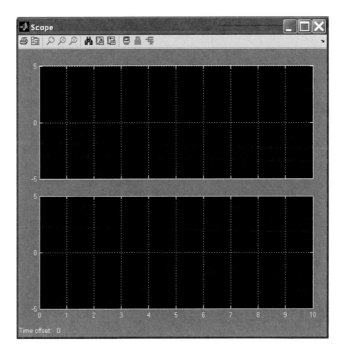

FIGURE 6.5 Two-input scope.

that it cannot continue, as we see from Figure 6.9. However, the simulator reading the message can determine quite easily that this system has the solution

$$x_k = \frac{2}{3} \text{ and } y_k = 5\frac{1}{3}.$$

What is the problem for the software?

Let's track how value propagation in the x_k layer will function in this model when we run the simulation. The value-propagation algorithm will first look for an input to the Scope in the x_k layer. To get the Scope's input, it will seek the Add block's output in this layer, but this will require that the algorithm find the inputs to the Add block. During the process of getting the Add block's inputs, the algorithm must cross between layers to get the input to the Scope in the y_k layer. But getting the input for the Scope in the y_k layer will ultimately bring the algorithm back to the input of the Scope in the x_k layer, causing the algorithm to repeat endlessly. While this is happening in the x_k layer, the same thing will be happening

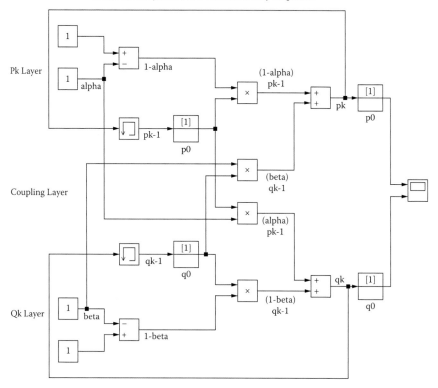

FIGURE 6.6   Final model for the panel survey opinion dynamics.

in the y_k layer. The circular paths prevent the algorithm from ever finding a value to propagate forward, and the simulation will repeat indefinitely.

To prevent this from happening, Simulink has an algorithm to detect such cases and stop them from running. It will print the message and give the simulator a chance to eliminate the circular path from the model. The simulator has control over this algorithm operation through the Configuration Parameters | Diagnostics window. The entry named Algebraic Loop must be set to error to stop the solver. If the simulator wants to ignore algebraic loops, this entry can be set to none, and the systems will continue to work on the solution. Interestingly, the solution algorithm can generate the correct answer for this algebraic loop if the entry is set to none.

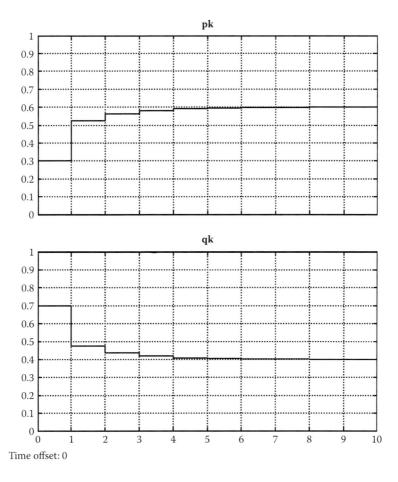

FIGURE 6.7 Output of panel survey opinion model. The *y*-axis displays the fraction in unitless values, and the *x*-axis displays the time step in units of interviews.

EXERCISE 6.1

The economic model of Exercise 3.2 was formed from the system of first-order difference equations for the relationship between supply and demand shown below

$$p_{n+1} = a - bq_{n+1} \tag{6.8}$$

$$q_{n+1} = \frac{1}{k} p_n \tag{6.9}$$

Model _6_2_1 Demonstration of an Algebraic Loop

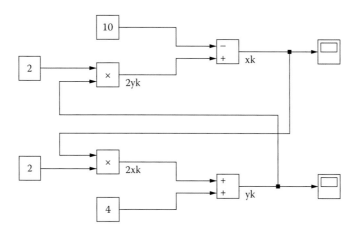

FIGURE 6.8   A model illustrating an algebraic loop.

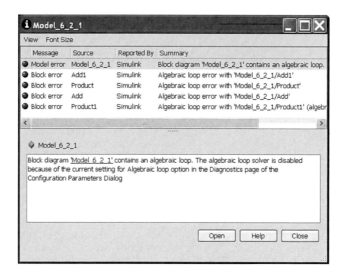

FIGURE 6.9   Error information for algebraic loop model.

where p_n is the price and q_n is the quantity of the goods at time n.

Create a Simulink simulation of the above system that shows the change in price for 20 days if the initial price is $p_0 = 30$, the constant $a = 50$, and

$$\frac{b}{k} \text{ is an input constant.}$$

Run the simulation for three values of

$$\frac{b}{k}: 0.9, 1.0, \text{ and } 1.1.$$

Assume that each day's price remains constant throughout the day and changes overnight.

6.4 WHAT IS A FIRST-ORDER DIFFERENTIAL EQUATION SYSTEM?

Just as in the case of difference equations, a system of first-order differential equations is just a set of first-order differential equations with interaction terms.

It is usually the case that the right-hand side of the equations contains terms that represent the effect of the other $y_i(t)$ on a particular $y(t)$. These terms are the means by which the solutions are coupled to each other.

A system of first-order differential equations is a set of n first-order, ordinary, differential equations describing the dynamics of the functions $y_1(t), y_2(t), \ldots, y_n(t)$ in the form

$$\dot{y}_1(t) = f_1(y_n(t), y_{n-1}(t), \ldots, y_1(t), t) \tag{6.10}$$

$$\dot{y}_2(t) = f_2(y_n(t), y_{n-1}(t), \ldots, y_1(t), t) \tag{6.11}$$

$$\ldots$$

$$\dot{y}_n(t) = fn(y_n(t), y_{n-1}(t), \ldots, y_1(t), t) \tag{6.12}$$

6.5 EXAMPLES OF FIRST-ORDER DIFFERENTIAL EQUATION SYSTEMS

A good example of a first-order system of differential equations comes from Edelstein-Keshet (1988). If a population of rabbits has a population of foxes as predators, the dynamics of the rabbit and fox populations can be modeled by the system of linear, first-order, ordinary differential Equations (6.13) and (6.14)

$$\frac{dR(t)}{dt} = g_R R(t) - d_{RF} F(t) \tag{6.13}$$

$$\frac{dR(t)}{dt} = g_{RR} R(t) - d_F F(t) \tag{6.14}$$

where $R(t)$ and $F(t)$ are the rabbit and fox populations in units of number of rabbits in the area and number of foxes in the area. The population-change terms on the RHS of the equations are g_R, the growth in rabbits due to reproduction and death, d_{RF}, the decrease in rabbits due to predation by foxes, g_{FR}, the growth in foxes due to predation of rabbits, and d_F, the decrease in foxes due to death.

The interaction term in Equation (6.13) is the term $-d_{RF}F(t)$, since this term gives the effect on the $R(t)$ solution due to $F(t)$. Similarly, the term $g_{FR}R(t)$ is the interaction term for Equation (6.14) giving the $F(t)$ solution. This pair of equations must be solved simultaneously for the results to be valid, since the instantaneous values of $R(t)$ and $F(t)$ affect both solutions.

6.6 SIMULATING A FIRST-ORDER DIFFERENTIAL EQUATION SYSTEM

In this section, we look at constructing a simulation of systems whose behavior is described by a system of coupled, first-order, ordinary differential equations. Since this is a system whose functions of interest are functions of each other's simulation, we must use a layered approach like that of the system of first-order difference equations. This means that we can construct our simulation by producing one layer of blocks in the block diagram for each first-order ODE (ordinary differential equation) in the system. Then we interconnect the outputs of the layers to the appropriate inputs in other layers through blocks representing the interactions

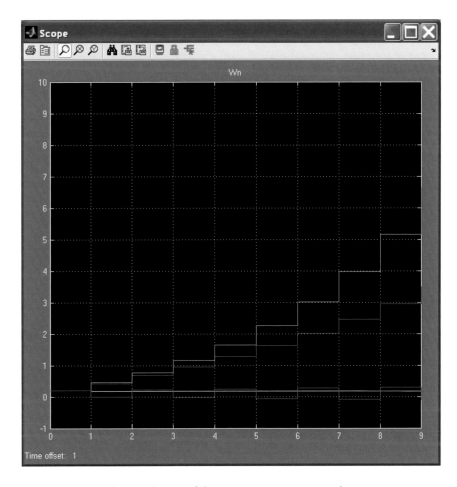

FIGURE 3.21B   The simulation of the vector-constant example.

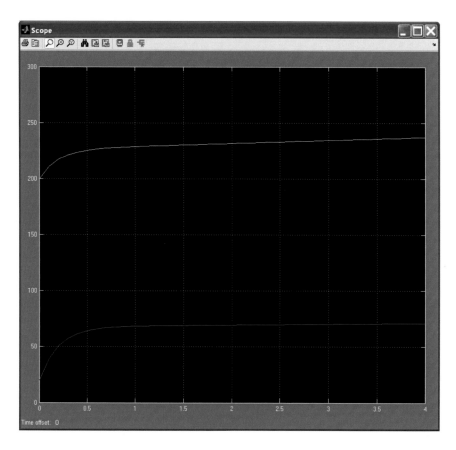

FIGURE 6.18   Output from the simulation of the rabbit–fox system using a signal bus.

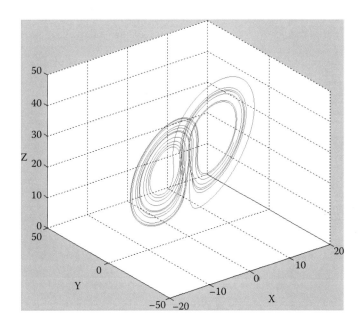

FIGURE 10.23   Lorenz model simulation results in 3-D Cartesian space.

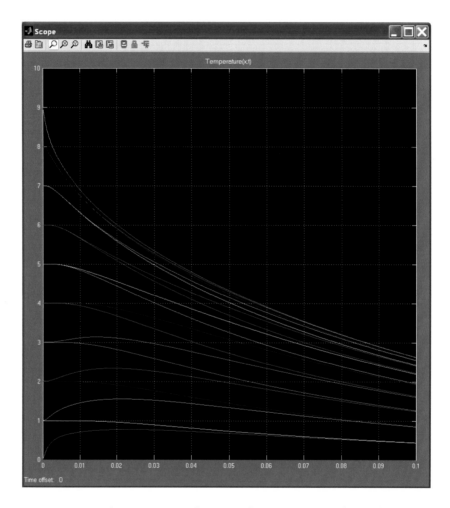

FIGURE 10.34   The temperature dynamics for 20 points in a thin rod.

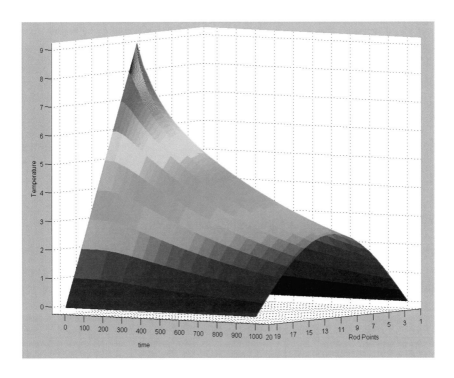

FIGURE 10.37   The temperature–time surface for the thin rod.

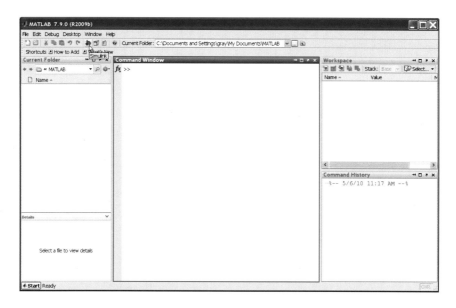

FIGURE B.3   Location of the Simulink icon on the MATLAB toolbar.

FIGURE C.10   The Debugger at the first breakpoint.

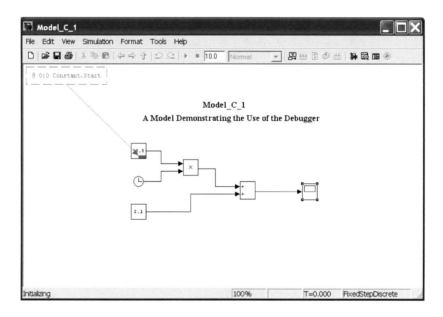

FIGURE C.12 The model window showing the breakpoint location.

between the functions. This kind of layered crosscoupling can work for systems of any number of equations.

As an example, we will develop a model for the previous rabbit-fox population model. An important way to approach the development of an interacting model is to produce the noninteracting version first and then add the interactions. So the first step is to construct a two-layer model for Equations (6.13) and (6.14) in our diagram without including the interaction terms. This model is shown in Figure 6.10.

Note that we have used the Subsystem block to simplify the visual aspects and to highlight the interaction terms when we include them. Figure 6.11 shows the details of the noninteracting subsystems. The right-hand subsystem is the Rabbit Population Model and the left-hand subsystem is the Fox Population Model.

To run the simulation for an area of 50 km², we take a starting population of 200 rabbits and 20 foxes. The values of the population-change constants can be estimated to be $g_R = 1.0$ year^{-1} and $d_F = 6.0$ year^{-1}. Using these values in Model_6_3_1 with a Fixed-Step ode4 solver with a time step of 0.1 gives the output in Figure 6.12. These results are exactly what we expect when one population grows and the other dies exponentially.

The next step is to modify the uncoupled, two layer Model_6_3_1 to include the interaction terms. To do this, we'll first modify the Subsystem blocks to include a Sum block to add in the effects of the cross-coupled

A Model of the Dynamics of a Rabbit-Fox Population

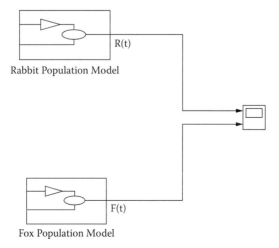

FIGURE 6.10   Model_6_3_1 for the rabbit–fox population system.

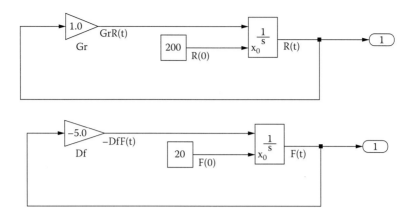

FIGURE 6.11   Subsystems of the rabbit–fox population Model 6_3_1.

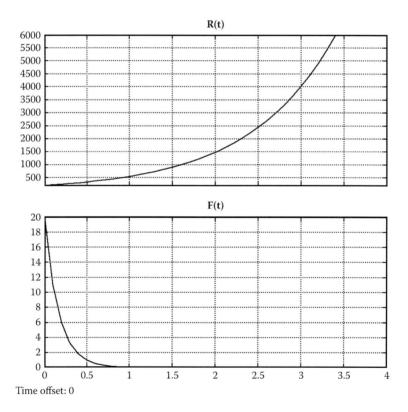

Time offset: 0

FIGURE 6.12   Results for the noninteracting rabbit–fox system. The *x*-axis shows the simulation time in years, and the *y*-axis shows the population of the species in individuals.

outputs, creating an input port for each subsystem. This input port will be how our interaction input will enter a layer. Then we connect the output of each layer into the input of the other layer's subsystem. Last, we estimate the interaction constants as $d_{RF} = 3.3$ year^{-1} and $g_{FR} = 1.8$ year^{-1}. This produces Model_6_3_2 shown in Figure 6.13. The Subsystem expansions are shown in Figure 6.14. The results of a simulation run for Model_6_3_2 are shown in Figure 6.15.

A Model of the Dynamics of a Rabbit-Fox Population

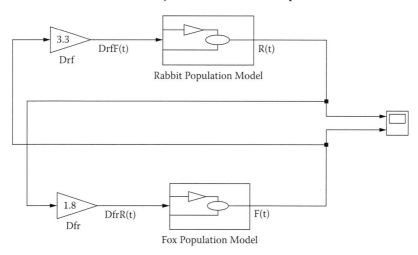

FIGURE 6.13 Model_6_3_2 for the interacting rabbit–fox system.

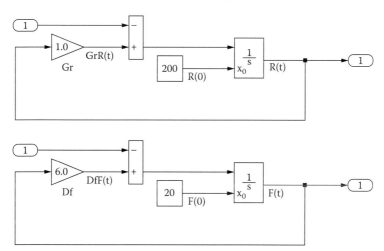

FIGURE 6.14 Expansion of subsystems for Model_6_3_2.

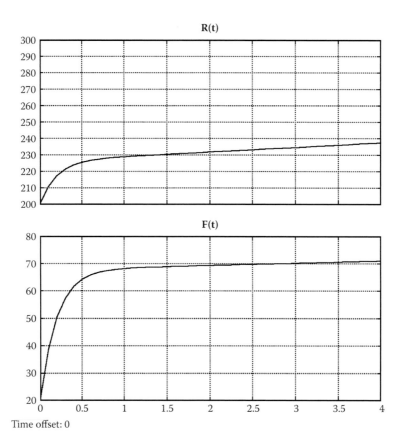

FIGURE 6.15 Results for the rabbit–fox interacting population system. The x-axis shows the simulation time in years, and the y-axis shows the population of the species in individuals.

We see from these curves that the predation of the foxes causes the system population to approach stabilization at about 70 foxes and about 240 rabbits over a 4-year period. The interaction between the two populations causes each to approach a stable population in this dynamic equilibrium.

6.7 COMBINING CONNECTIONS ON A BUS

We often need to collect a set of connections into a single entity in our simulation diagrams. We can meet this need by introducing a new block into our simulations.

6.7.1 The Bus Creator Block from the Signals Routing Library

The Bus Creator block from the Signal Routing library is a way to bundle connections into a single line to reduce visual complexity when the signals are routed through a complex diagram. This bundle of connections is represented as if it were a vector signal, with each individual signal as a component.

To use a Bus Creator block, we must set the block's Number of Inputs ports parameter on the Block Parameters screen, shown in Figure 6.16, to the number of connections we want to bundle over the bus. The block then displays input ports equal to the number we specified, and we can connect the desired signals to the resulting input ports.

We can discriminate among the signals in the bus by a name that the Bus Creator assigns. There are two bus signal naming options. We can specify that each signal on the bus inherit the name of the signal or that each input signal have a user-assigned name. The Bus Creator generates its own names for unnamed bus signals by using "signal n," where n is the number of the input port to which the signal is connected. There is a list on a Bus Creator's parameter window that displays all the signals on the bus. A plus sign (+) next to a signal indicates that the signal is a bus and can be expanded.

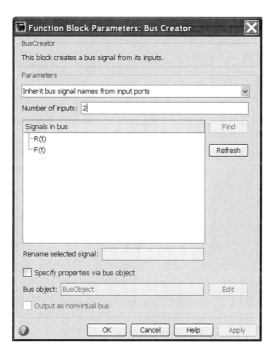

FIGURE 6.16   The Bus Creator block parameters.

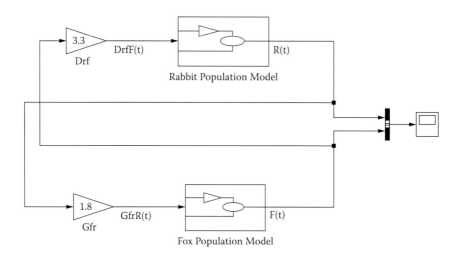

FIGURE 6.17   Model_6_3_3 for the rabbit–fox system using a signal bus.

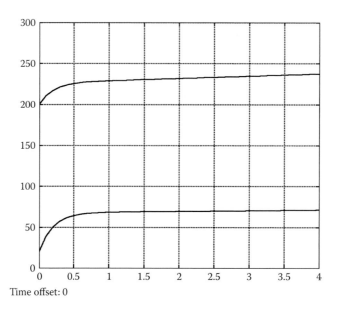

Time offset: 0

FIGURE 6.18   (See color insert following page 144.) Output from the simulation of the rabbit–fox system using a signal bus. The x-axis shows the simulation time in years, and the y-axis shows the population of the species in individuals.

What we will do is use the Bus Creator output to display our rabbit and fox populations on the same Scope. First we'll disconnect the R(t) and F(t) signals from the inputs to the Scope. Then we'll insert a Bus Creator block into the diagram and reconnect the signals into the Bus Creator block (2 signals is the default number of input lines). Finally, we'll change the number of inputs to the Scope to 1 and connect the output of the Bus Creator block to the Scope. This produces Model_6_3_3, shown in Figure 6.17. Running the simulation again produces the output shown in Figure 6.18.

Since the Bus Creator output is represented as a vector, the input to the Scope is a vector, and it will assign a different color to each component in the order described in Table 3.6. This color assignment helps to distinguish the signals from each other (see Chapter 3). In the case of Figure 6.18, the top line is the rabbit population, R(t), and is yellow, while the lower line is the fox population, F(t), and is magenta.

EXERCISE 6.2

Simulate the system described by the set of first-order equations

$$\dot{X}(t) = 0.5\,\text{min}^{-1}\,X(t) - 0.5\,\text{min}^{-1}\,Z(t) \quad X(0) = 1 \qquad (6.15)$$

$$\dot{Y}(t) = 0.5\,\text{min}^{-1}\,X(t) \quad Y(0) = 1 \qquad (6.16)$$

$$\dot{Z}(t) = 0.05\,\text{min}^{-1}\,Y(t) \quad Z(0) = 0.79 \qquad (6.17)$$

with a fixed step size of 0.01 min for the range 0.0 to 12.0 min, using a *Runge–Kutta* algorithm. Use a Bus Creator block to gather the outputs and display them on a single-input Scope. What are the values of $X(12.0)$, $Y(12.0)$, and $Z(12.0)$ to an accuracy of two decimal places?

EXERCISE 6.3

Keen and Spain (1992) describe the Chance–Cleland model for enzyme–substrate interaction that assumes the reaction

$$E + S \underset{k_2}{\overset{k_1}{\rightleftarrows}} C \xrightarrow{k_3} E + P \qquad (6.18)$$

In this reaction, E is the enzyme, S is the substrate, C is the enzyme–substrate compound, and P is the product. The rate constants for the reactions are k_1, k_2, and k_3. The dynamic equations for the kinetics of the reactions are:

$$\frac{d[S]}{dt} = -k_1[S][E] + k_2[C] \tag{6.19}$$

$$\frac{d[C]}{dt} = -k_1[S][E] - (k_2 + k_3)[C] \tag{6.20}$$

$$\frac{d[E]}{dt} = -k_1[S][E] - (k_2 + k_3)[C] \tag{6.21}$$

$$\frac{d[P]}{dt} = k_3[C] \tag{6.22}$$

where $[E]$ is the enzyme concentration, $[S]$ is the substrate concentration, $[C]$ is the enzyme–substrate compound concentration, and $[P]$ is the product concentration.

Create a Simulink simulation that shows the change in concentration of these four reactants when the initial concentrations are $[S] = 100$, $[E] = 10$, $[C] = 0$, $[P] = 0$, and the rate constants are $k_1 = 0.005$ s^{-1}, $k_2 = 0.005$ s^{-1}, and $k_3 = 0.1$ s^{-1}. Run the simulation from 0 to 200 s.

6.8 SUMMARY

In this chapter, we have seen systems of first-order equations serving as models. A system of first-order differential equations is a set of n first-order, ordinary differential equations (ODE) describing the dynamics of the system. The RHS of the system of equations contains terms that represent the effect of the other $y_i(t)''$ on a particular $y_j(t)$ and are the means by which the solutions are coupled to each other.

These systems can be organized into models using layers that interconnect through interaction terms. Simulations are constructed by producing one layer of blocks in the block diagram for each first-order ODE in the system. The outputs of layers are connected to the inputs of other layers through blocks representing the interactions between the functions.

Algebraic loops can be created in a Simulink model by feedbacks that cause the current inputs of a block to depend on its current outputs. The `Bus Creator` block from the Signal Routing library is a way to bundle signals into a single line to reduce visual complexity, and its output is a `Vector` of the input signals.

REFERNCES AND ADDITIONAL READING

Artzrouni, M., and J. Gouteux. 2000. A Model for the Spread of Sleeping Sickness. In *Applied Mathematical Modeling: A Multidisciplinary Approach.* Ed. D. Shier and K. Wallenius. Boca Raton, FL: Chapman & Hall/CRC.

Dym, C. L. 2004. *Principles of Mathematical Modeling.* New York: Elsevier.

Edelstein-Keshet, L. 1988. *Mathematical Models in Biology.* New York: McGraw-Hill.

Goldberg, S. 1986. *Introduction to Difference Equations.* New York: Dover.

Haberman, R. 1977. *Mathematical Models: Mechanical Vibrations, Population Dynamics, and Traffic Flow.* Englewood Cliffs, NJ: Prentice Hall.

Huckfeldt, R., C. Kohfeld, and T. Likens. 1982. *Dynamic Modeling: An Introduction.* Beverly Hills, CA: Sage.

Keen, R. E., and J. Spain. 1992. *Computer Simulation in Biology: A Basic Introduction.* New York: John Wiley.

Mooney, D., and R. Swift. 1999. *A Course in Mathematical Modeling.* Washington, DC: Mathematical Association of America.

Severance, F. L. 2001. *System Modeling and Simulation: An Introduction.* New York: John Wiley.

The MathWorks, Inc. 2007. http://www.mathworks.com.

CHAPTER 7

Simulation of Second-Order Equation Models

Nonperiodic Dynamics

THE DYNAMICS OF MANY systems are described by second-order equations, and they are probably more common among those analyzed by scientists and engineers than any other kind. We now consider how we can create simulations of systems whose dynamical equations are of second order. We begin with second-order difference equation models and then move on to second-order differential equation models. We also concentrate on systems whose motion is nonperiodic.

7.1 SIMULATION OF SECOND-ORDER DIFFERENCE EQUATION MODELS

We recall that a difference equation is first order when the expression on the right-hand side (RHS) of the defining equation contains terms in time step $n-1$ only. The following is the equation of a second-order difference equation:

$$y_n = f(y_{n-1}, y_{n-2}, n) \tag{7.1}$$

We found, for first-order difference equations, that a simulation is easily constructed as a block diagram, provided that we have a way of using y_{n-1} at each time step n. Investigation showed us that we could produce the value for y_{n-1} by sampling the output at n and delaying it one time step by storing the value in a Memory block. After the time step, we used the stored value to compute the next value. An IC (Initial Condition) block can be used to produce the initial value for y_{n-1}.

But how can we use a Memory block to store a value *two* time steps in the past? The answer is to use *two* Memory/IC blocks so that each Memory/IC block combination produces the value delayed by another time step. This way, we can simulate any second-order difference equation simply by having two block combinations.

As an example of the construction of the two-block combination, consider the following second-order equation:

$$y_n = y_{n-1} + 2y_{n-2} + n \qquad \text{with } n = 2, \ldots, 5 \text{ and } y_0 = 1, y_1 = 2 \qquad (7.2)$$

At this point, we want to translate Equation (7.2) into a simulation structure using a pair of Memory/IC block combinations. On consideration, we realize that there are two ways to arrange the combinations of the simulation into a structure. Both kinds of structure have merit, so we consider each in the following discussion.

7.1.1 Sequential Structure

The first of our two possible structures is a sequential structure that is the logical extension of the structure we chose for the first-order models. To see how we do this, let's reverse the equation and group the terms according to the time step they relate to. We'll also separate them horizontally by time-step term.

$$\left[y_{n-1} \right]_{\text{layer } n-1 \text{ terms}} + \left[2y_{n-2} \right]_{\text{layer } n-2 \text{ terms}} + \left[n \right]_{\text{layer } n \text{ terms}} \rightarrow y_n \qquad (7.3)$$

We see that this equation looks like the first-order equations we simulated earlier, but this equation has an additional term on the left-hand side. So all we need to do to extend our first-order structure is to add an additional term to the structure. This indicates that we need to have one Memory/IC block combination for the $n-1$ terms on the leftmost part of the left-hand side. Then we must have another Memory/IC block combination for the $n-2$ terms in the middle part. The n output must feed back to supply the $n-1$ terms with the value to store for the next time step, while the $n-1$ terms must feed their output back to the $n-2$ terms with their storage value. Putting all this into our diagram, we construct Model_7_1_1 shown in Figure 7.1.

To run the simulation, we must set the Fixed-step Discrete solver and choose a time step of 1. Since the model is only defined for 2 to 5, we must set the start time to 2 and the end time to 5. Running the model under these condition gives the output shown in Figure 7.2.

Model_7_1 Sequential Structure for Second-Order Difference Equation Model Simulation

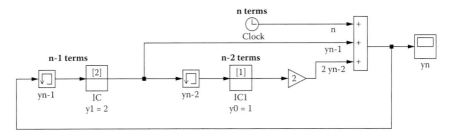

FIGURE 7.1   Model_7_1_1: sequential structure of a second-order difference equation simulation.

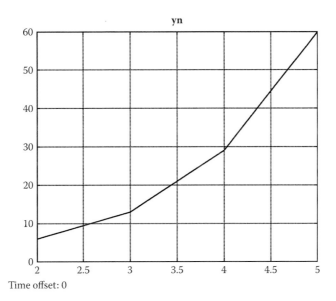

FIGURE 7.2   The output of the sequential structure for Model_7_1_1. The simulation time is shown on the x-axis, and the model output on the y-axis, both in arbitrary units.

7.1.2 Layered Structure

Turning to our second possible structure, let's reverse Equation (7.2) and group the terms according to the time step they relate to, just as we did in the sequential structure. But this time we'll separate them vertically by time-step term.

$$+\left[n\right]_{\text{layer } n \text{ terms}} \rightarrow y_n$$

$$+\left[y_{n-1}\right]_{\text{layer } n-1 \text{ terms}} \qquad (7.4)$$

$$+\left[2y_{n-2}\right]_{\text{layer } n-2 \text{ terms}}$$

At this point, we want to translate Equation (7.3) into a simulation structure. The arrangement here suggests a block diagram with three subsystems to represent the three time-step terms. Figure 7.3 shows a simulation structure with three subsystems that correctly simulates the example second-order Equation (7.2) and illustrates how the layering is formed. Each layer is marked to show how the outputs of the layers are combined

Model_7_1_2 Layered Structure for Second-Order Difference Equation Model Simulation

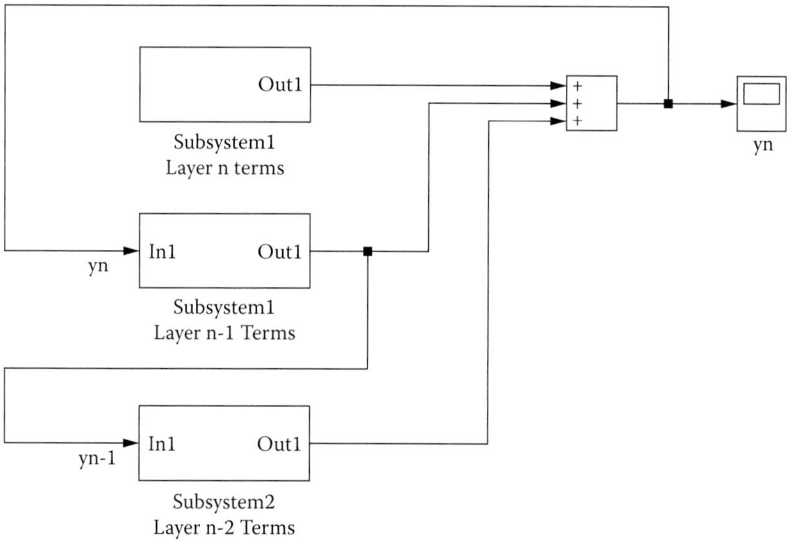

FIGURE 7.3 Layered structure of a second-order difference equation simulation for Model_7_1_2.

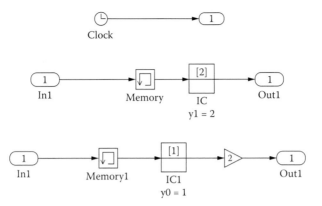

FIGURE 7.4 Layer details for the layered structure for Model_7_1_2.

to contribute to the total output as well as propagate from layer to layer to save preceding values, as shown in Model_7_1_2.

In this structure, the top layer provides the output for the part of the right-hand side that depends on time step n. Since the only function in this layer is n, it only contains a Clock block providing us with the value for n, as shown in Figure 7.4.

The second layer provides the output for the part of the right-hand side that depends on time step $n-1$. In this layer, we must have a Memory block to provide the output for the function y_{n-1} term after the initial time step and the initial value $y_1 = 2$ for the initial time step. The third layer contains a Memory block to provide the y_{n-2} term and the initial value $y_0 = 1$ for the initial time step. The layered structure causes the output of the second layer, y_{n-1}, to be stored in the third layer, thus providing y_{n-2} in the next time step. Note that we must be careful to feed back the output of the y_{n-1} layer *after* the IC block so that we store the initial $y_1 = 2$ value for use in computing the y_{n-2} term in the next time step. (We don't need to save the $y_0 = 1$ value because it's only used in the first time step.) Note also that we've used a Gain block for the factor 2 in the term $2y_{n-2}$ instead of using a Product block in the y_{n-2} layer. This is done to save space and make a smaller diagram.

If we run this model using the same configuration parameters as the sequential structure, we find an identical output, as we expect. These structures are equivalent, and the choice of which to use should be made to suit the application. A large complex simulation with many interacting subsystems would probably be best served by using the layered structure.

A small simulation of limited elements would probably be fastest to construct using the sequential structure.

EXERCISE 7.1

Simulate the system described by the linear, second-order difference equation

$$y_n = 3y_{n-1} - 2y_{n-2} + \frac{1}{n} \quad \text{with } n = 2,\ldots,8 \quad \text{and } y_0 = 2,\, y_1 = 2 \tag{7.5}$$

EXERCISE 7.2

Gersting (1993) describes a model of a bank account savings accumulation. We can adapt her model to consider a bank that wants to encourage long-term savings account creation by adopting an interest payment strategy that will encourage initial depositors to keep long-term deposits. For the first two years, an account receives a flat 3% on the total balance at the end of each year. Beginning at the end of the third year of an account's lifetime, the account is awarded 5% interest on the total balance at the end of that year, 2% on the total balance at the end of the previous year, and 1% on the total balance at the end of the year before that.

An account is established with $1,000 at the start of a year. In addition to the interest received, the account holder makes an additional deposit of $500 midway through each year.

1. Create a Simulink® simulation using IC/Memory blocks that shows the growth of this account over a 10-year period.
2. Create a Simulink simulation that shows the growth of an account earning a flat 3% interest over a 10-year period.
3. Determine which of these accounts would produce the highest interest for the account holder.

7.2 SIMULATION OF SECOND-ORDER DIFFERENTIAL EQUATION MODELS

Second-order ordinary differential models are those whose equations are in the form

$$\ddot{y}(t) + a(t)\dot{y}(t) + b(t)y(t) = f(y(t),t) \tag{7.6}$$

Turning to the simulation of these models, we first ask how we can systematically construct equivalent block diagrams. When we dealt with first-order equations, our approach was to write the dynamical equation by isolating the highest-order derivative term on the left-hand side and putting the other terms on the right-hand side. Then we integrated the left-hand side to produce the function needed for output. The expression on the right-hand side became the diagram supplying the output value of the function. We use this same technique on our second-order equation.

Moving all terms except the highest-order derivative to the right-hand side and renaming constants so that there are no minus signs gives

$$\ddot{y}(t) = c(t)\dot{y}(t) + d(t)y(t) + f\big(y(t),t\big) \tag{7.7}$$

Let us now perform an integration of both sides from t_0 to t.

$$\int_{t_0}^{t} \ddot{y}(t)dt = \int_{t_0}^{t}\big[c(t)\dot{y}(t) + d(t)y(t) + f\big(y(t),t\big)\big]dtw \tag{7.8}$$

$$\int_{t_0}^{t} \frac{d}{dt}\dot{y}(t)dt = \int_{t_0}^{t}\big[c(t)\dot{y}(t) + d(t)y(t) + f\big(y(t),t\big)\big]dt \tag{7.9}$$

$$\dot{y}(t)\big|_{t=t} - \dot{y}(t)\big|_{t=t_0} = \int_{t_0}^{t}\big[c(t)\dot{y}(t) + d(t)y(t) + f\big(y(t),t\big)\big]dt \tag{7.10}$$

$$\dot{y}(t) = \dot{y}(t_0) + \int_{t_0}^{t}\big[c(t)\dot{y}(t) + d(t)y(t) + f\big(y(t),t\big)\big]dt \tag{7.11}$$

We perform a second integration of both sides to find

$$\int_{t_0}^{t} \frac{d}{dt}y(t)dt = \int_{t_0}^{t}\Big[\dot{y}(t_0) + \int_{t_0}^{t}\big[c(t)\dot{y}(t) + d(t)y(t) + f\big(y(t),t\big)\big]dt\Big]dt \tag{7.12}$$

$$y(t)\big|_{t=t} - y(t)\big|_{t=t_0} = \int_{t_0}^{t}\left[\dot{y}(t_0) + \int_{t_0}^{t}\left[c(t)\dot{y}(t) + d(t)y(t) + f(y(t),t)\right]dt\right]dt \tag{7.13}$$

$$y(t) = y(t_0) + \int_{t_0}^{t}\left[\dot{y}(t_0) + \int_{t_0}^{t}\left[c(t)\dot{y}(t) + d(t)y(t) + f(y(t),t)\right]dt\right]dt \tag{7.14}$$

This is the form of the solution that we need to construct our simulation. Note that we need *two* initial values to solve this equation. This is consistent with the requirement that an *n*th-order differential equation requires *n* constants to specify the solution uniquely.

Let us rearrange this equation and spread it out vertically so we can see the structure more clearly.

$$y(t_0) + \int_{t_0}^{t}\Big[\qquad\qquad\qquad\qquad \Big]dt \rightarrow y(t)$$

$$\uparrow$$

$$\dot{y}(t_0) + \int_{t_0}^{t}\Big[\qquad\qquad\qquad \Big]dt$$

$$\uparrow$$

$$c(t)\dot{y}(t) + d(t)y(t) + f(y(t),t) \tag{7.15}$$

As in the difference equation case above, this arrangement can be translated into two different diagrams, sequential structure and layered structure.

7.2.1 Sequential Structure

To construct a sequential structure for this equation, we arrange the lines in Equation (7.15) from left to right. The output of the simulation is on the right side of our diagram, and it is the term $y(t)$ from the top line. It is the output of an Integrator block with an initial value $y(t_0)$ and an input produced by an integrator computing the integral in the second line. The output of the second-line integrator is produced from an initial value $\dot{y}(t_0)$ and the output of a subnetwork computing the function on the last line. The inputs to the function are the products $c(t)\dot{y}(t)$, $d(t)y(t)$, and $f(y(t),t)$. The term $c(t)\dot{y}(t)$ is the output of a Product block with inputs $c(t)$ and $\dot{y}(t)$. The term $d(t)y(t)$ is the output of a Product block with inputs $d(t)$ and $y(t)$. The last term, $f(y(t),t)$, is the output of a network that simulates the right-hand side with input $y(t)$. Note that, in this second-order model, we must have *two* feedback lines. The output of the second-line integrator provides the feedback to supply the input $\dot{y}(t)$, while the output of the top-line integrator provides the feedback to supply the input $y(t)$. Using this method, we can now construct the Simulink model in Figure 7.5.

As an example of a second-order differential equation system, consider the dynamical equations for an object falling from rest in a uniform, constant gravitational force field. The dynamical equations for this system are

$$\frac{d^2}{dt^2} y(t) = \frac{1}{m} F(y,t) \tag{7.16}$$

$$\frac{d^2 x(t)}{dt^2} = 0 \tag{7.17}$$

These equations are recognized as the two-dimensional equivalent of Newton's laws applied to an object of mass m in a force field directed along the y-axis. To complete the dynamical specification, we must supply two

Sequential Structure for Simulating Second-Order Differential Equations

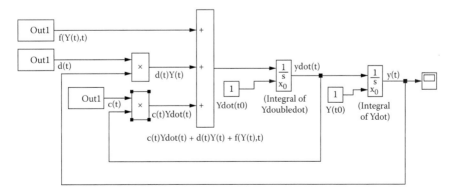

FIGURE 7.5 Sequential structure for a second-order differential equation model.

initial values for the equation: the mass of the object and the strength of the uniform, vertical force:

$$y(0)=100\text{m}, \quad \dot{y}(0)=0\text{m/s}, \quad m=2kg, \quad F(y,t)=-mG \tag{7.18}$$

$$x(0)=(0)\text{m}, \quad \dot{x}(0)=0\text{m/s} \tag{7.19}$$

In these equations, G, the gravitational constant of the uniform field, is 9.8 m/s². Note that the sign of the force is negative, indicating that the direction of the force points downward (positive y to negative y).

Using the general structure from Equation (7.15), we can construct a sequential structure simulation, giving Model_7_2_1 in Figure 7.6. Note that we have used both a scope and an XY graph for output. We use two scopes to show the values for $y(t)$ and $\dot{y}(t)$ and the XY graph to show the object trajectory in the x–y plane (x and y).

To run this model, we choose a Fixed-step solver of the continuous type. Let's choose ode3, the solver that implements the Euler method, and run the simulation for 10 s with a time step of 0.1 s. Figure 7.7 shows the output we produce from this simulation.

Note several things about these results. First, the y position of the object changes in a parabolic fashion, although the x position stays constant (the object falls vertically downward). Next, the velocity of the object increases linearly downward (negatively), which is exactly what we expect for a uniformly accelerating object. The object shows no signs of reaching a terminal value, and this is consistent with a model in which there are no limiting terms such as viscosity. In a vacuum, there is no terminal velocity for an object falling in a force field. Finally, the time it takes to reach the ground ($y = 0$) from 100 m up is about 4.5 s.

7.2.2 Layered Structure

Let's return to Equation (7.15) and see how we can construct a layered structure. Two of the layers in this case will contain Integrator blocks and their initial conditions, and there will be a third layer for the $c(t)\dot{y}(t)+d(t)y(t)+f(y(t),t)$ function. The top layer will be the Integrator receiving $\dot{y}(t)$ as an input and producing $y(t)$ as an output. The middle layer will be the Integrator receiving as an input the result of computing $c(t)\dot{y}(t)+d(t)y(t)+f(y(t),t)$ and producing $\dot{y}(t)$ as its output. The bottom

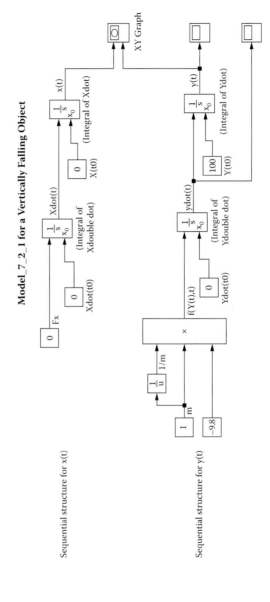

FIGURE 7.6 Sequential structure for Model_7_2_1.

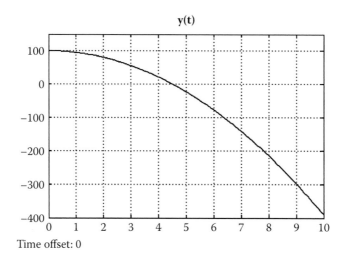

Time offset: 0

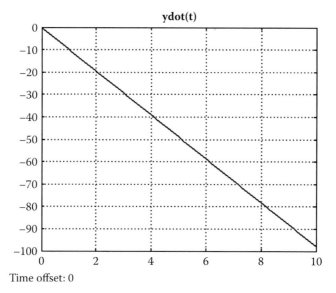

Time offset: 0

FIGURE 7.7 The results of Model_7_2_1 simulation. The height in meters is given on the *y*-axis, and the time in seconds is shown on the *x*-axis. (*Continued*)

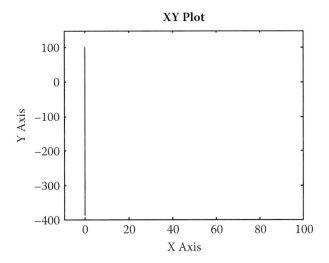

FIGURE 7.7 (*Continued*)   The results of Model_7_2_1 simulation.

layer will compute the function $c(t)\dot{y}(t)+d(t)y(t)+f(y(t),t)$. We expect two feedback lines going to the bottom layer, one to supply $\dot{y}(t)$ from the middle layer and one to supply $y(t)$ from the top layer. Figure 7.8 shows the layered structure, and Figure 7.9 shows the subsystem details.

Using our example of the object falling in the uniform gravitational field, we can construct an example layered model. Figure 7.10 shows the final model with three subsystems for the three layers. Figure 7.11 shows the expansion of the three layers for the $y(t)$ component.

EXERCISE 7.3

Simulate the motion of a pebble thrown upward from the surface of the Earth into the atmosphere. Assume that its initial velocity is 50 m/s. The simulation should provide

1. The height to which the pebble rises
2. The speed of the pebble when it hits the ground
3. The total time between the release of the pebble and the impact with the ground

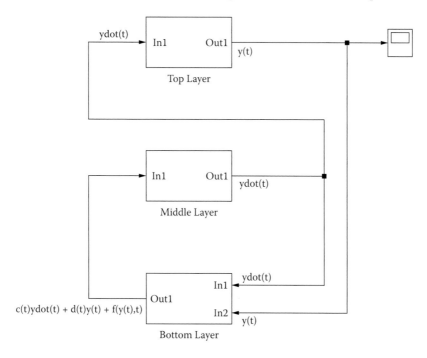

Layered Structure for Simulating Second-Order Differential Equations

FIGURE 7.8   Layered structure for a second-order differential equation model.

7.3 SECOND-ORDER DIFFERENTIAL EQUATION MODELS WITH FIRST-ORDER TERMS

In the previous section, we considered an example system whose dynamics were described by a second-order differential equation with no first-order terms. In this section, we examine how we simulate second-order models that do have first-order terms. First-order terms in second-order differential equations are *dissipative* terms that allow friction and other similar dissipative effects to be included in our models.

The simulation serving as our example is the object falling in a uniform gravitational field from the previous section. We'll add a viscous fluid exerting a frictional effect to the model, and this will produce an example of first-order terms. We will use the sequential structure.

7.3.1 Viscosity Modeled by a Linear Function of Velocity

If the object is falling in a viscous medium such as a gas or liquid, one physical theory of the viscous force acting on the falling object is that the viscosity adds an *opposing* force of resistance. In this model, we take the

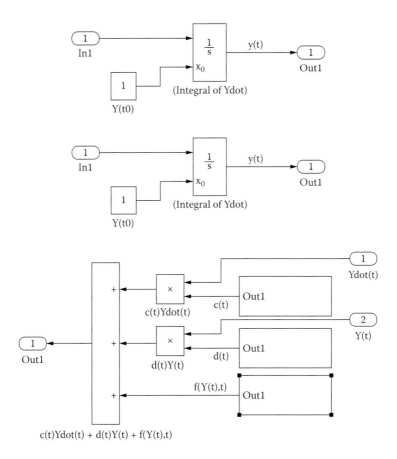

FIGURE 7.9   Subsystem expansions for the layered structure for a second-order differential equation model.

viscosity to be a linear function of the object's velocity and model it as a linear first-order term, $k_1 \dot{y}(t)$ that always acts in the opposite direction to the direction of the object's velocity. The strength of the viscosity is determined by the constant k_1, which has units of s^{-1}. Including this model of the effects of air resistance produces the dynamical equation for the object falling in a viscous medium and uniform gravitational field

$$\ddot{y}(t) = k_1 \dot{y}(t) - G \quad \text{with } \dot{y}(0) = 0\text{m/s}, \, y(0) = 100\text{m}, \, k_1 = 4.8\text{s}^{-1} \quad (7.20)$$

(The value of the linear viscosity constant, k_1, is an empirically determined value.) The sequential simulation for this equation is Model_7_3_1 and is

Model_7_2_2 for a Vertically Falling Object

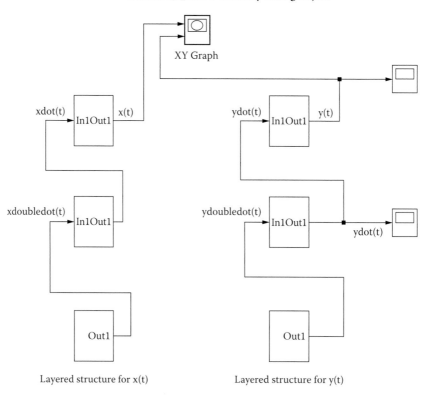

Layered structure for x(t)   Layered structure for y(t)

FIGURE 7.10   Layered structure for Model_7_2_2 for the falling object.

shown in Figure 7.12. In this model, we are representing the downward force of the gravitational field by one subsystem and the upward force of the viscous resistance by another subsystem. The subsystem for the viscosity resistance is shown in Figure 7.13.

Note that the velocity, $\dot{y}(t)$, is fed back into the viscosity subsystem to enable it to produce the linear viscosity force. The Gain block of −1 is necessary to reverse the sign of the negative velocity and make it oppose the gravitational force. The results with this model are given in Figure 7.14.

Note that the results change substantially when we include the linear first-order term. The velocity no longer continues to increase without bound. Instead, there is a terminal velocity of almost 30 m/s for the given viscosity constant. This is consistent with our observations of objects falling near the Earth's surface. The time at which the object strikes the surface, $y = 0$, is now lengthened to about 6 s. Again, this is reasonable for a

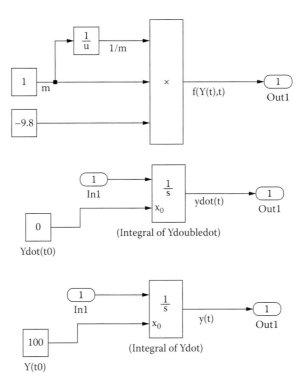

FIGURE 7.11 $y(t)$ subsystem expansions for Model_7_2_2.

viscous medium, since the viscosity acts to buoy up the object as it falls, and it takes longer to reach the surface.

The models for resistance are usually empirically determined, so other theories are also feasible. In particular, some theories of viscous resistance are quadratic, rather than linear, and they are found to model some situations better than the linear model.

7.3.2 Viscosity Modeled by a Quadratic Function of Velocity

Changing the simulation to use a quadratic model $k_2 \dot{y}^2(t)$ is easy and is a vivid demonstration of the power of simulation. No new theoretical analysis is necessary, since the simulation structure remains the same. We merely replace the linear viscosity subsystem with a quadratic viscosity subsystem and recompute the results. Even more significantly, this simple change causes the dynamical equation to become a *nonlinear* second-order differential equation, making the theoretical solution even more difficult.

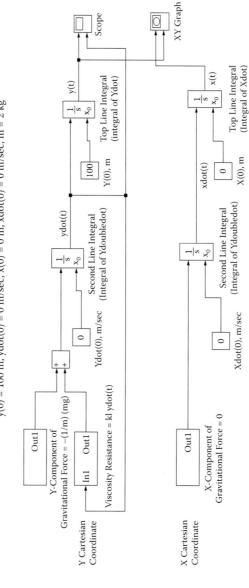

FIGURE 7.12 Simulation for an object falling in a viscous fluid for Model_7_3_1.

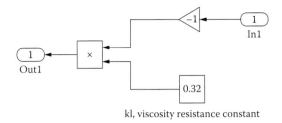

FIGURE 7.13 Linear viscosity force subsystem for Model_7_3_1.

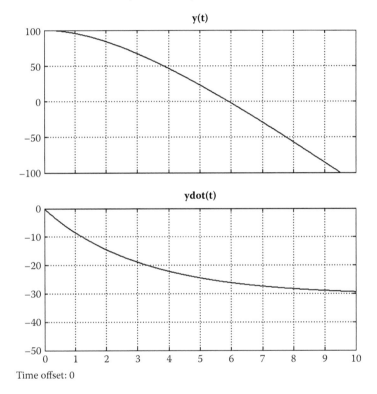

Time offset: 0

FIGURE 7.14 Results of the simulation of Model_7_3_1 for an object falling in a viscous medium. The height in meters is given on the y-axis, and the time in seconds is shown on the x-axis. *(Continued)*

The new subsystem for Model_7_3_2 is shown in Figure 7.15. Note that the Gain block has been removed in this subsystem, since squaring the negative velocity produces a positive term, which is necessary to oppose the gravitational force and is a simple modification of the linear subsystem. We have used the square function to produce the $\dot{y}^2(t)$ term. When

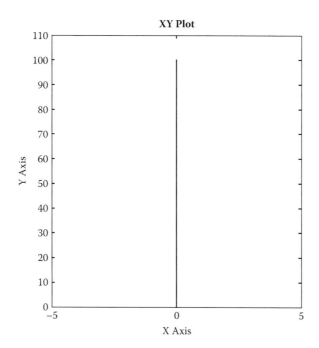

FIGURE 7.14 (*Continued*)   Results of the simulation of Model_7_3_1 for an object falling in a viscous medium.

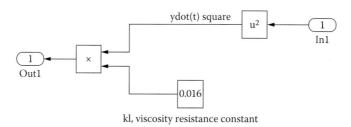

FIGURE 7.15   Quadratic viscosity force subsystem for Model_7_3_2.

we make this change, we also need to use a different empirically determined constant of 0.16 m^{-1}.

The results produced by the quadratic model are shown in Figure 7.16. These results are similar to the results for the linear viscosity model, although the quadratic model reaches terminal velocity somewhat faster than the linear model. Experimentation is needed to determine which model fits reality better.

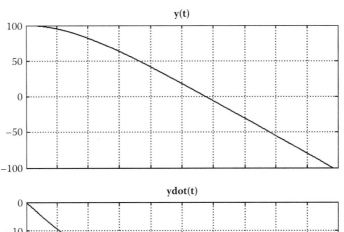

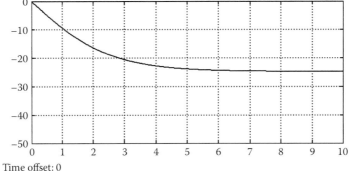

Time offset: 0

FIGURE 7.16 Results of the simulation of Model_7_3_2 for a quadratic viscous medium. The height in meters is given on the y-axis, and the time in seconds is shown on the x-axis. *(Continued)*

EXERCISE 7.4

Simulate the motion of a pebble dropped from a height of 100 m above the surface of the Earth in the atmosphere. Assume that the drag force exerted by the atmosphere is proportional to v_y^2 and that the value of the terminal speed for the pebble has been experimentally determined to be 30 m/s (Hint: adjust the viscosity constant k_2 until the simulation terminal speed matches the experimentally observed value.) The simulation should provide

1. The speed of the pebble when it reaches 90 m above the surface of the Earth
2. The total time between the release of the pebble and the impact with the ground

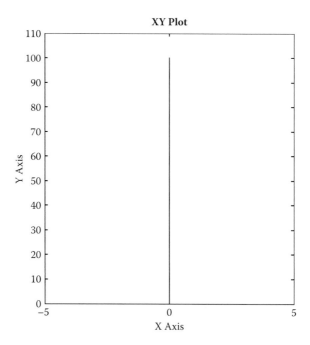

FIGURE 7.16 (*Continued*)   Results of the simulation of Model_7_3_2 for a quadratic viscous medium.

7.4 CONDITIONAL DYNAMICS

Our models of objects falling in gravitational fields have been somewhat unrealistic, since expanding the size of our scopes and graphs shows that the objects continue to move in the negative direction even after they have impacted the ground at $y = 0$. This aggravation can be removed from our model by using *conditional blocks*. In this section we will see how these blocks work and use them to make our simulations more realistic.

7.4.1 Object Moving at the Earth's Surface

Let's first generalize our model of an object moving in a uniform gravitational field so that our model will accommodate an object traveling at an arbitrary angle θ to the horizontal direction. In physics, scientists have found that the model for this motion can be broken into two independent parts: motion along the y-axis and motion along the x-axis. Since the gravitational force only acts in the y-direction, the x-axis motion is unaffected by the gravitational field. This analysis gives rise to the dynamical equations

$$\frac{d^2}{dt^2} x(t) = 0 \qquad\qquad (7.21)$$

$$\frac{d^2}{dt^2} y(t) = \frac{1}{m}(-mG) \qquad\qquad (7.22)$$

which must be supplemented with the values

$$\dot{y}(0) = v_0 \sin\theta \quad m/s$$

$$y(0) = y_0 \quad m$$

$$\dot{x}(0) = v_0 \cos\theta \quad m/s$$

$$x(0) = x_0 \quad m$$

$$G = -9.8 \quad m/s^2$$

Putting this dynamical information into the sequential model Model_7_4_1 shown in Figure 7.17—using an angle of 60°, an initial velocity of 20 m/s, initial positions of y = 100 m and x = 0 m, and then running this model for 10.0 s using a fixed-step ode3 solver with a time step 0.01 s—gives the results shown in Figure 7.18 for the x–y trajectory of the object.

EXERCISE 7.5

Resimulate the motion of a pebble from Exercise 7.4 assuming that it is thrown upward at an angle of 30° from the surface of the Earth into the atmosphere with an initial velocity of 50 m/s. Assume that the drag force exerted by the atmosphere is proportional to v_y^2 and that the value of the terminal speed for the pebble has been experimentally determined to be 30 m/s. (Use the value of k_2 that was found in Exercise 7.4.) The simulation should provide

1. The speed of the pebble when it reaches 90 m above the surface of the Earth
2. The height to which the pebble rises
3. The total time between the release of the pebble and the impact with the ground

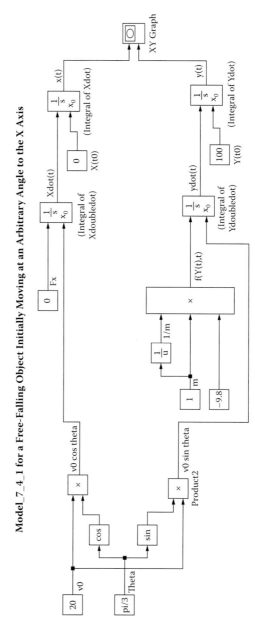

FIGURE 7.17 Model_7_4_1 for an object moving at an arbitrary angle in the Earth's field.

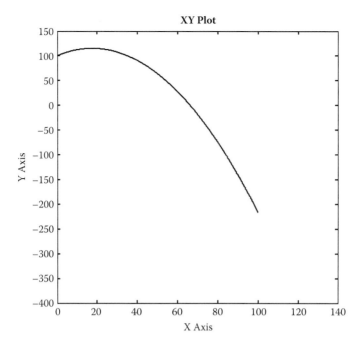

FIGURE 7.18 Trajectory results for the object of Model_7_4_1. The height in meters is given on the y-axis, and the time in seconds is shown on the x-axis.

Examining these results shows that the object does not stop when it impacts the ground ($y = 0$). We would like to modify this model so that it shows the correct behavior when the object reaches the ground. That means when $y = 0$, the velocity of the object in both directions drops to 0 and the value of x remains fixed at its value when $y = 0$ is reached. A way to describe these actions is to use the logical statements shown in Equations (7.23) and (7.24).

$$\text{If } y(t) \geq 0 \text{ then output } y(t) \text{ else output } 0 \qquad (7.23)$$

$$\text{If } y(t) \geq 0 \text{ then output } x(t) \text{ else output x at the time when } y(t) = 0$$
$$(7.24)$$

What we need to do is to use an appropriate block or set of blocks to implement these two statements. We deal with the statement in Equation (7.23) first, since it is the easiest.

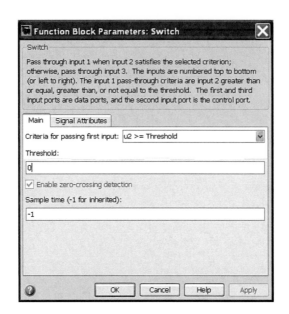

FIGURE 7.19 The Switch block parameters.

7.4.1.1 The Switch Block from the Signal Routing Library

The Switch block from the Signal Routing library is exactly what we need for the Equation (7.23) implementation. This block receives three inputs; a middle input serves as a control that causes either the top or the bottom input to be routed to the block output. The Switch block has the parameters shown in Figure 7.19.

From this, we see that we can set a numerical condition to be met by the value of the control input. If the condition is met, the *top* input is routed to the block output. Otherwise, the bottom input is routed out. Our control condition can be $y \geq 0$, in which case we route $y(t)$ through the top input to the output. Otherwise, we route 0 through the bottom input to the output. A modification that provides the appropriate dynamics for the $y(t)$ output is shown in Figure 7.20.

EXERCISE 7.6

The Collatz function (Stanoyevitch 2005) is a recursive function of a positive, nonzero integer that is conjectured to settle into the cycle of values, 1, 4, 2, for any initial integer. It is defined by

$$i_{n+1} = i_n/2 \quad i_n \text{ even} \quad i_0 \geq 1, n \geq 0 \qquad (7.25)$$

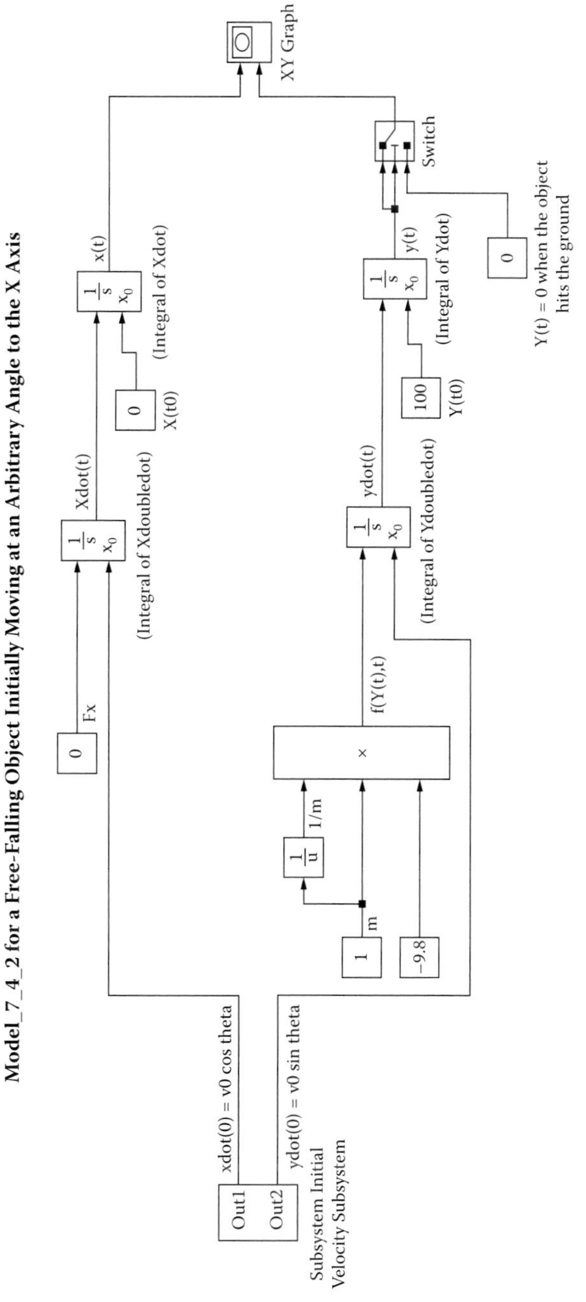

Model_7_4_2 for a Free-Falling Object Initially Moving at an Arbitrary Angle to the X Axis

FIGURE 7.20 Modification of the $y(t)$ output line for Model_7_4_2.

$$i_{n+1} = 3i_n + 1 \quad i_n \text{ odd} \quad i_0 \geq 1, n \geq 0 \qquad (7.26)$$

Simulate the output of the Collatz function by constructing a difference equation model that uses a Switch block for outputting the values of this function.

Turning to the Equation (7.24) implementation, we see that this task is more complex than the Equation (7.23) task, because the value of $y(t)$ will always be 0 when the pebble hits the ground, while the value of $x(t)$ will depend on the initial velocity in the x-direction. To do this task, we must sample the value of $x(t)$ when the pebble hits the ground and then save that for outputting during the remainder of the simulation. Examining the libraries, we find that three blocks are necessary.

7.4.1.2 The Zero-Order Hold Block from the Discrete Library

First, we must have a block that samples its input at a particular time to form its output, and then it must continue to output that value for a stated length of time. The Zero-Order Hold block has exactly this behavior. Its input is sampled and then output for the length of time specified in its Sample time parameter, as shown in Figure 7.21.

If the value inf is entered into this field, the block will sample its input once and hold it forever. This is what we need for the $x(t)$ output. But we still need to control the time at which the Zero-Order Hold block is allowed to sample the $x(t)$ value. The next block that we study allows us to control when the Zero-Order Hold will operate.

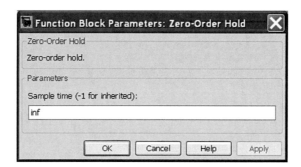

FIGURE 7.21 The Zero-Order Hold block parameters.

7.4.1.3 The If Action Subsystem from the Ports & Subsystems Library

The If Action subsystem suppresses operation of its components until it receives an enabling signal on its Action port. If we insert our Zero-Order Hold block in an If Action subsystem, it will not sample its input until it receives a signal, even though it is connected to a functioning input. If we can arrange an enabling signal to be sent to the If Action subsystem when the $y = 0$ condition occurs, then the Zero-Order Hold will be activated, sample its input, and provide as its continuing output the $x(t)$ signal when $y = 0$ occurs. The last block we need is the block to provide the If Action subsystem with its enabling signal.

7.4.1.4 The If Block from the Ports & Signals Library

The If block is designed to test an input signal and activate an enabling signal on the output port associated with the condition. This block can have multiple outputs, each with its own condition, so it acts as a kind of controller for a set of If Action subsystems. The parameters associated with this block are shown in Figure 7.22.

Note that multiple output ports can be created by entering multiple conditions in the Elseif expressions parameter and that the default settings provide an Else output port. We do not need either of these features, so we clear the Show else condition setting. Finally, we must set the If expression to our condition, which is $y \leq 0$.

7.4.1.5 The Merge Block from the Signal Routing Library

Having created two signals representing one output, we must merge these two signals into one so that we can connect it to the XY Graph input. The Merge block from the Signal Routing Library makes this possible. Its parameters are shown in Figure 7.23.

Returning to our correction of the $x(t)$ line in the simulation, we combine all these elements into a single subsystem, shown in Figure 7.24, and place it between the last Integrator and the Scope. Figure 7.25 shows the final diagram incorporating all of the modifications given above, culminating in Model_7_4_3. The results of running this simulation are shown in Figure 7.26.

The simulation has the correct behavior: The object stops when it reaches ground at the $y = 0$ line, just as we would expect in reality. If we use a Floating Scope block to examine the behavior of $y(t)$ and $x(t)$ separately, we find two curves. The solid line is the $y(t)$ curve after conditional modification. This curve mimics the unmodified $y(t)$ until it reaches the

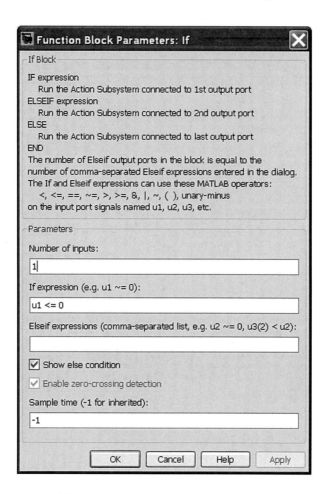

FIGURE 7.22   The If block parameters.

ground at $y = 0$, after which it becomes a horizontal straight line. The dashed straight line that increases linearly and then levels off is the function $x(t)$ after modification. It reaches its maximum value just at the point $y = 0$, after which it remains at that value.

EXERCISE 7.7

Resimulate the motion of a pebble from Exercise 7.5 with a realistic model that shows the pebble stopping when it hits the surface. Be sure to include an XY Graph block that shows this behavior.

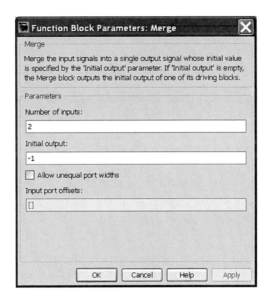

FIGURE 7.23   The Merge block parameters.

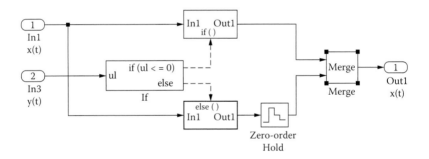

FIGURE 7.24   The $x(t)$ correction subsystem for the realistic simulation.

EXERCISE 7.8

Use a simulation to determine the optimum angle to throw a shot in the shot-put event to achieve maximum range. Assume that the initial height of the shot is 2 m and that the shot is thrown with an initial velocity of 30 m/s. Assume that the air resistance can be modeled by a viscous force k_2v^2, where k_2/m for the shot is 0.001. Make sure that the shot stops when it hits the ground.

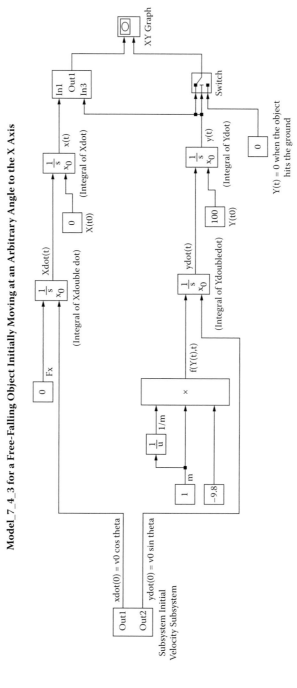

Model_7_4_3 for a Free-Falling Object Initially Moving at an Arbitrary Angle to the X Axis

FIGURE 7.25 Model_7_4_3 for the realistic simulation of an object falling onto the Earth's surface.

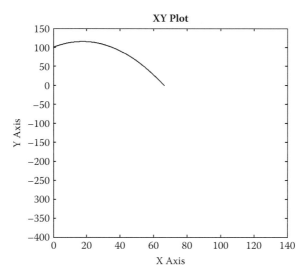

FIGURE 7.26 Results of the realistic simulation of Model_7_4_3. The height in meters is given on the *y*-axis, and the time in seconds is shown on the *x*-axis.

7.5 SUMMARY

Second-order models are commonly found in nature, so the simulation of these models fills an important need. We use sequences or layers to organize our models.

The direct simulation of a higher-order differential equation is achieved by a simulation layer with a sequence of integrators. The number of integrators must be equal to the order of the highest-order derivative. First-order terms in second-order differential equations are dissipative terms for the system being modeled

Conditional blocks can be used to model systems with abrupt changes in behavior that are triggered by conditions. The Switch block, If block, and If Action block are used to provide conditional behavior. The Zero-Order Hold block samples and holds its input for a specified sample period.

REFERENCES AND ADDITIONAL READING

Duan, J. 2000. Modeling Nonlinear Phenomena by Dynamical Systems. In *Applied Mathematical Modeling: A Multidisciplinary Approach*. Ed. D. Shier and K. Wallenius, 231–240. Boca Raton, FL: Chapman & Hall/CRC.

Dym, C. 2004. *Principles of Mathematical Modeling*. New York: Elsevier Academic.

Haberman, R. 1977. *Mathematical Models: Mechanical Vibrations, Population Dynamics, and Traffic Flow*. Englewood Cliffs, NJ: Prentice Hall.

Garcia, A. 2000. *Numerical Methods for Physics*. Upper Saddle River, NJ: Prentice Hall.

Gould, H., and J. Tobochnik. 1996. *An Introduction to Computer Simulation Methods: Applications to Physical Systems*. New York: Addison Wesley.

Higham D. J., and N. Higham. 2005. *MATLAB Guide*. Philadelphia: Society for Industrial and Applied Mathematics.

Klee, H. 2007. *Simulation of Dynamic Systems with MATLAB and Simulink*. Boca Raton, FL: CRC Press.

Knight, R. D., B. Jones, and S. Field. 2007. *College Physics: A Strategic Approach*. New York: Pearson Addison Wesley.

CHAPTER 8

Simulation of Second-Order Equation Models

Periodic Dynamics

IN THIS CHAPTER, WE continue our exploration of second-order models, the most important models in science and engineering, by examining orbital systems that serve as good examples of systems with periodic dynamics.

8.1 ORBITAL SYSTEMS

These result from *central force interactions*—those kinds of forces that are exerted along the straight line connecting two bodies, as shown in Figure 8.1. There are a variety of these kinds of forces, and analysis of their dynamics leads to second-order differential equations expressing the accelerations of the objects in the systems due to the central forces. The gravitational field is an excellent example of a central-force system.

8.1.1 The Gravitational Attraction between Two Objects

When two objects with mass in space approach each other, they react to a gravitational force between them. The force is given by the equation

$$\bar{F} = -\frac{GMm}{r^3}\bar{r} \tag{8.1}$$

where M is the mass of one object, m is the mass of the other object, $\bar{F}$ is the force vector acting on each object, $\bar{r}$ is the radius vector between the

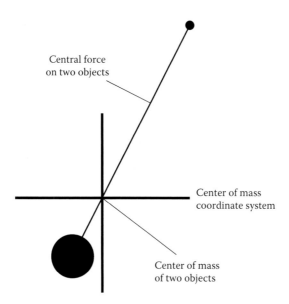

FIGURE 8.1 Coordinate system for the two-body, central-force system.

objects, and G is the gravitational constant, 6.67×10^{-11} m³/kg × s². The objects will move about the center of mass of the two objects, which is located on the radius vector between them.

8.1.2 The Earth–Sun System

An important case of such a system is the Earth–Sun system. In this case, the mass of the Sun, M, is so much larger than the Earth that the center of mass can be taken to be the center of the Sun. This means that we can set our coordinate system at the center of the Sun and regard the Sun as stationary throughout the simulation.

An analysis of the central-force law in Equation 8.1 applied to the Earth–Sun system leads to dynamical equations for the position of the Earth as a function of time

$$\ddot{y}(t) = -\frac{GM}{r^3} y(t) \qquad \dot{y}(0) = v_0^y, \, y(0) = y_0 \tag{8.2}$$

$$\ddot{x}(t) = -\frac{GM}{r^3} x(t) \qquad \dot{x}(0) = v_0^x, \, x(0) = x_0 \tag{8.3}$$

where $r^2 = x^2 + y^2$ and $v^2 = v_x^2 + v_y^2$. Since the motion of the Earth about the Sun due to the central force is planar, we only need to simulate this system in two dimensions. We use the Cartesian coordinate system, and the equations are written in terms of the x and y coordinates of the Earth. Note especially that each equation is coupled to the other through the radius, r.

There are two important properties of interest in the physics of this system: the total energy and the z-component of the total angular momentum. These are

$$E = \frac{1}{2}mv^2 - \frac{GMm}{r} \tag{8.4}$$

$$L_z = m(xv_y - yv_x) \tag{8.5}$$

and these must be conserved (that is, held constant) throughout the dynamical behavior of the system.

Recall from Chapter 2 our discussion of the need to choose the units of the simulation so that numerical values are comfortably located in the range of computable values, and a simulation using astronomical units provides an excellent choice. These units replace the very large and small values for distance and the gravitational constant in the SI system with values that are more suitable for our numerical simulation. First, we define the AU (astronomical unit) to be the semimajor distance from the Sun to the Earth (the Earth's orbit is an ellipse), 1.496×10^{11} meters. Next, we take the units of time to be a year, 3.15×10^7 seconds.

When we do this, we realize that the distance from the Earth to the Sun is very close to 1 AU at aphelion and the period of the Earth around the Sun is 1 year. The product of the gravitational constant and the Sun's mass, GM, becomes

$$4\Pi^2 \frac{\text{AU}^3}{\text{yr}^2}$$

(Gould 1996). We also choose units for the energy and angular momentum relative to the Earth's mass by using energy and angular momentum

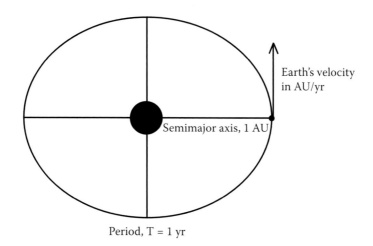

Earth's velocity in AU/yr

Semimajor axis, 1 AU

Period, T = 1 yr

FIGURE 8.2   Earth–Sun geometry in astronomical units.

absolute units divided by the Earth's mass. Then the Earth–Sun geometry becomes that shown in Figure 8.2.

These units and the preceding analysis lead us to the simulation of a model of two coupled, second-order equations in a sequential structure, as seen in Figure 8.3. In this model, note that there are three subsystems: one providing two factors involving the gravitational constant, GM/r^3 and GM/r, a second providing the angular momentum, and a third the total energy. These subsystems receive the current values for the position coordinates and the component velocities from the sequential structures implementing the dynamical equations.

The results of running this simulation are shown in Figure 8.4 demonstrating the velocity and position as a function of time for both coordinates, the total energy and z-component of the angular momentum as a function of time, and the Earth's trajectory around the Sun (taken to be located at 0,0).

Note that the x- and y-positions are not pure sinusoidal functions, since the orbit is an ellipse, and that the total energy and z-component of the angular moment are constant, as required.

8.1.3 Circular Orbits

Could the Earth have coalesced into a *circular* orbit instead of the elliptical one we have? For this to be the case, the velocity of the Earth would have had to satisfy the relation

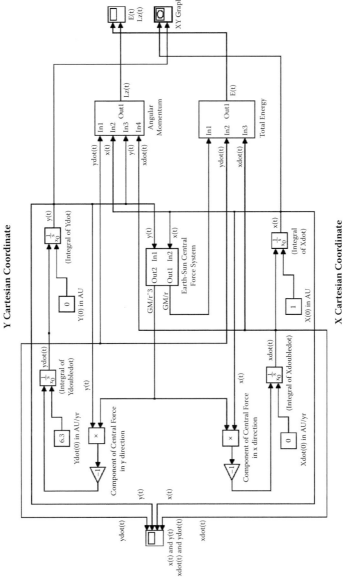
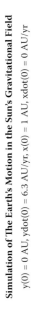

Simulation of The Earth's Motion in the Sun's Gravitational Field

y(0) = 0 AU, ydot(0) = 6.3 AU/yr, x(0) = 1 AU, xdot(0) = 0 AU/yr

FIGURE 8.3   Model_8_1_1 for the simulation of the Earth–Sun system.

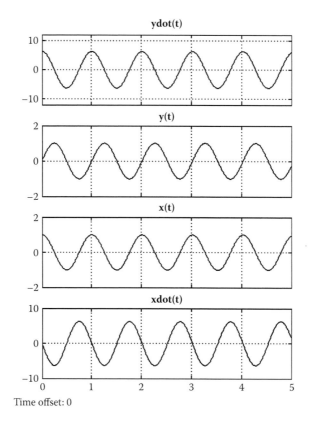

FIGURE 8.4   Results of running the simulation of Model_8_1_1. The leftmost graphs are the velocity and position coordinates in units of AU and AU/yr. The middle graphs show the Earth's energy and z-component of the angular momentum in relative units. The rightmost graph is the trajectory of the Earth in AU units.   *(Continued)*

$$V_{\text{Earth}} = \sqrt{\frac{GM}{r}} \tag{8.6}$$

We can investigate this question by seeing if we can achieve a circular orbit in our simulation. We incorporate the condition from Equation (8.6) in our model by setting the initial velocity such that it satisfies Equation (8.6) using an output from the Force subsystem, as shown in Model_8_2_1 in Figure 8.5. All we needed to do was to take Model_8_1_1 and replace the externally supplied constant initial condition by an internally supplied input from the square root of the *GM/r* output from the Force subsystem. The results from this model are shown in Figure 8.6. Note that the velocity

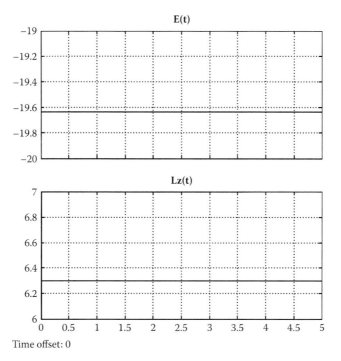

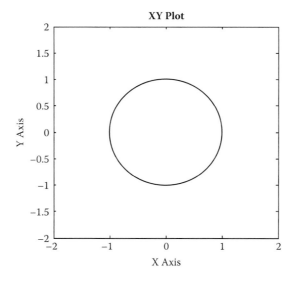

FIGURE 8.4 (*Continued*)   Results of running the simulation of Model_8_1_1. The leftmost graphs are the velocity and position coordinates in units of AU and AU/ yr. The middle graphs show the Earth's energy and z-component of the angular momentum in relative units. The rightmost graph is the trajectory of the Earth in AU units.

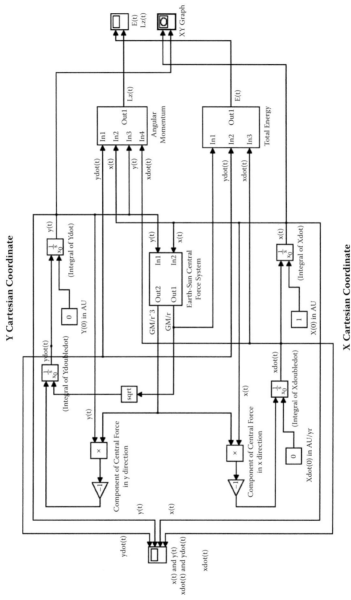

FIGURE 8.5   Model_8_2_1 for Earth to have a circular orbit.

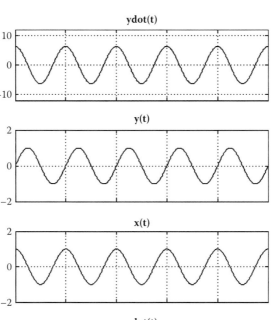

FIGURE 8.6 Results of the circular Earth-orbit simulation. The leftmost graphs show the velocity components and position coordinates in units of AU and AU/yr. The rightmost graph shows the trajectory about the Sun in AU. *(Continued)*

and position functions are now sinusoidal functions, as we would expect from a circular orbit.

8.1.4 The Earth–Satellite System

It is obvious that the simulation we developed for the Earth–Sun system can be applied without structural change to an Earth–satellite system. But the units used for the Earth–Sun system are not good choices for the satellite problem. So we use the earth units (EU) system for this problem, rather than the astronomical units (AU) system. In the EU system, the unit of distance is the radius of the Earth, 6.37×10^6 meters. Time is measured in hours.

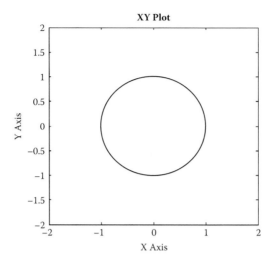

FIGURE 8.6 (*Continued*) Results of the circular Earth-orbit simulation. The left-most graphs show the velocity components and position coordinates in units of AU and AU/yr. The rightmost graph shows the trajectory about the Sun in AU.

The equations of motion for the satellite about the Earth are the same as Equations (8.2) and (8.3), except that the large mass, *M*, is now the mass of the Earth and the small mass, *m*, is that of the satellite. In the EU system, the gravitational constant *G* is 3.34×10^{-24} EU^3/kg × hr^2. Using this value for *G* gives a value of 20.0 EU^3/hr^2 (Gould 1996) for the product *GM*.

We can use the previous model with only changes to the constants to simulate the motion of a satellite in a circular, synchronous, equatorial orbit about the Earth. Recall that a synchronous orbit about the Earth has a period of 24 hours, so that the satellite remains above the same point on the Earth's surface at all times. We pick a starting position *x*(0) for the satellite so that it completes an orbit in exactly 24 hours. In our simulation, this value turns out to be *x*(0) = 6.6 EU, which produces an orbit radius that is slightly larger than the accurate value of 35,786 km, as seen in Figure 8.7. Running this model gives the results in Figure 8.8.

EXERCISE 8.1

Simulate the effect of drag resistance on a satellite orbiting the Earth. Assume that the viscous force is proportional to the square of the speed of the satellite with the magnitude of the viscous force equal to approximately 0.1 of the gravitational force. The satellite should be started in a

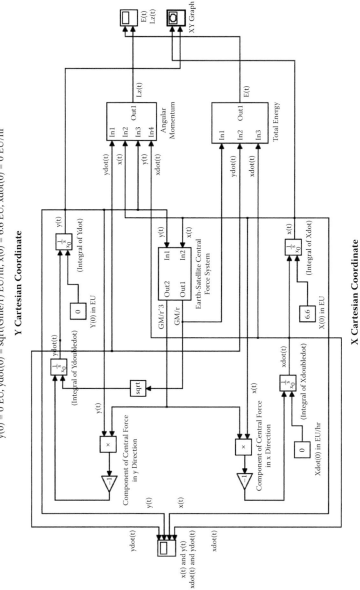

FIGURE 8.7 Model_8_3_1 for the Earth–satellite system.

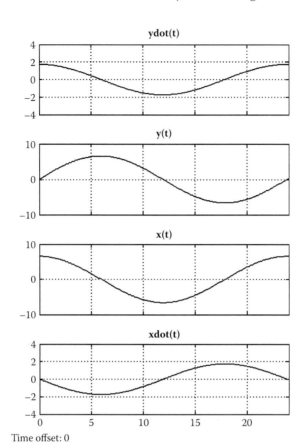

FIGURE 8.8 The results of the Earth–satellite simulation. The leftmost graphs show the velocity components and position coordinates in units of EU and EU/hr. The rightmost graph shows the trajectory about the Earth in EU. *(Continued)*

circular orbit and the effect of the drag should not begin until one orbit has completed.

8.2 MASKED SUBSYSTEMS

We have seen the value of using subsystems in a complex model with many input constants so that we can organize and lessen visual complexity. But, to change the constants of the model so that different alternatives can be explored, we must open the subsystems and change the constants manually. And if we want to adjust the parameters of other blocks being used, we must also change their values as well. Simulink® provides a mechanism to eliminate this clumsy procedure by providing a user-defined interface

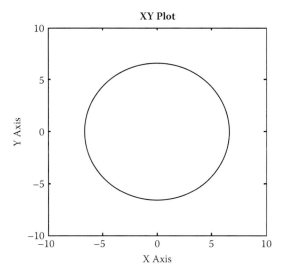

FIGURE 8.8 (*Continued*) The results of the Earth–satellite simulation. The left-most graphs show the velocity components and position coordinates in units of EU and EU/hr. The rightmost graph shows the trajectory about the Earth in EU.

to a user-defined subsystem. This interface is called a *mask* in Simulink terminology. In this section, we'll see how to create such an interface.

8.2.1 A Simple Example

The model in Figure 8.9 is a simple simulation of a linear function of its inputs. The linear function is determined by two constants: the y-axis intercept, b, and the slope of the line, m. Both of these are supplied as constants in the model.

The result of running this simulation with $b = 1$ and $m = 2$ is shown in Figure 8.10. Note that the y-axis intercept is 1 and the slope is 2, as expected.

We would now like to make this simulation into a subsystem to see how a mask can be created for the function. First we create a subsystem for the

Simple Linear Model Demonstrating a Masked Subsystem

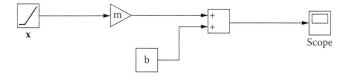

FIGURE 8.9   Model_8_4_1 demonstrating a masked subsystem.

three blocks in the model. This is accomplished by selecting the blocks and using the Edit | Create Subsystem command. This produces the model shown in Figure 8.11.

The three-block structure has been replaced by the subsystem with a default icon in the center. If we expand the subsystem, we find the model shown in Figure 8.12.

Now we turn to the creation of a mask for this subsystem so that the slope and intercept can be supplied by a user without requiring editing of the subsystem components.

8.2.2 Parameterizing the Subsystem Components

So far in this text, we have not used anything other than numerical constants as parameters for Constant and Gain blocks. To make the mask feasible, we must use *variable names* as our Constant and Gain block parameters. Simulink allows the use of variable names in

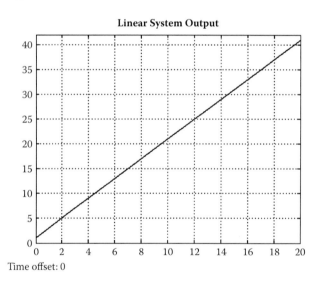

FIGURE 8.10   Simulation results for Model_8_4_1.

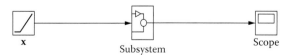

FIGURE 8.11   Model_8_4_2 for the masked subsystem demonstration.

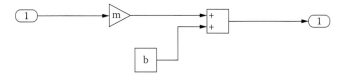

FIGURE 8.12   Subsystem details from Model_8_4_2.

parameter expressions. These variables are defined within the Simulink model workspace.

When a model is created or loaded within Simulink, a workspace for the model is created, and variables can be defined within this workspace. When the model is saved, the workspace is saved with it, and it is restored when the model is reloaded. The restrictions on variable-name uniqueness are the same in the workspace as in Simulink generally.

Using a variable name is easy; we just open the parameter block for the block and enter variable names that are not in use, as seen in Figure 8.13.

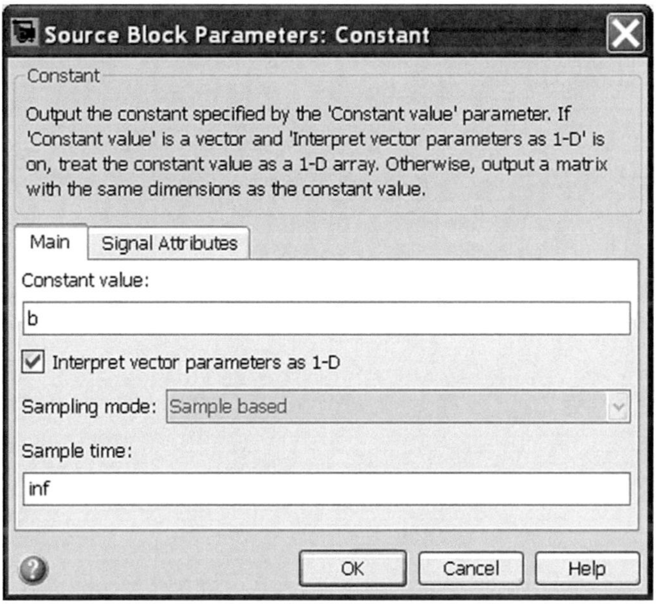

FIGURE 8.13   Using variable names in Constant and Gain block parameters.

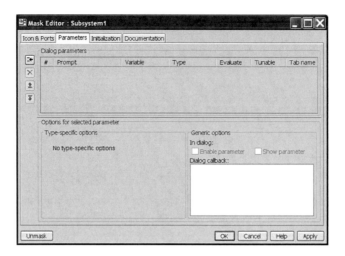

FIGURE 8.14 The Mask Editor.

8.2.3 Creating the Subsystem Mask

To create a mask for a subsystem, we must select the subsystem first, and then use the Edit | Mask Subsystem command. If we do this, we see the Mask Editor window in Figure 8.14. Note that this is a tabbed editor window and that we can assign an icon, parameters, initialization expressions, and documentation for the subsystem. In Figure 8.14, the Parameters tab has been brought to the foreground for creation of the mask parameters for the subsystem. The four buttons on the left side are for inserting, deleting, and moving parameters within the mask. The last button allows us to assign the order in which the parameters are seen by the user.

If we insert two new parameters—one for the slope m and one for the intercept b—we can define their prompts and their types. These two entries will define the parameters for the example subsystem.

Once parameters are defined for a subsystem, users who double-click the subsystem icon will now see a Block Parameters window just as if it were a built-in Simulink block. Then it can be edited just as any other block. This is the mask for the subsystem. Figure 8.15 shows the new Parameter Block window for our example subsystem.

For this mask, the `Documentation` tab of the Mask Editor window is used to provide a brief explanation of the subsystem for the user. The `Prompt` fields of the Parameters tab are used to label the parameters so that the user knows which one is entered. And the `Type` field of the `Parameters` tab has been used to define the parameter input as `Editable` so that the user

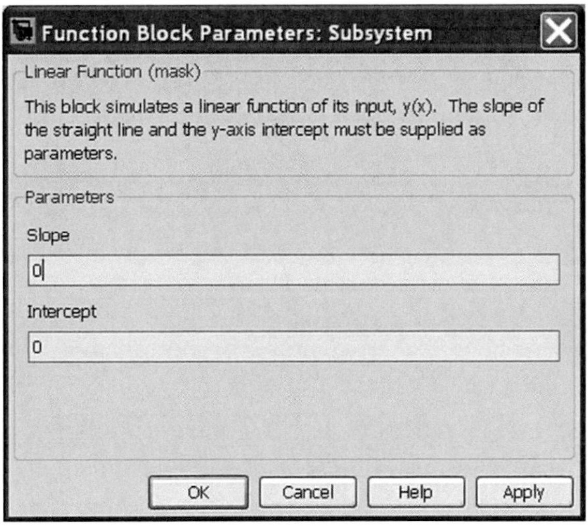

FIGURE 8.15 The example subsystem parameters.

can enter the needed value. Initialization fields allow a default value to be supplied so that the user can simply accept the defaults.

We can also edit the Icons tab so that we have a customized icon for our subsystem. This requires us to enter MATLAB® plotting functions so that Simulink can produce the icon by evaluating these functions. We have used the expression `plot([0 1],[0 0.5]);` to produce an icon consisting of a simple diagonal line. More examples of icon production can be found in the Simulink documentation. When we have finished, we have the masked subsystem shown in Figure 8.16.

EXERCISE 8.2

Consider a satellite in a circular orbit about the Earth at a radius of 3 EU.

1. Simulate the effect of a *radial* square impulse force exerted by an orbit-changing engine that fires with a force of 1 G (1 gravitational force) for 1 hour at 9.125 hours into the orbit, then turns off. Use a masked subsystem named *Square Impulse* to provide the square impulse function that controls the firing. *Square Impulse* must allow a user to enter the start time and stop time for the firing period as parameters without editing the components of the subsystem. Show the orbit for the satel-

(a)

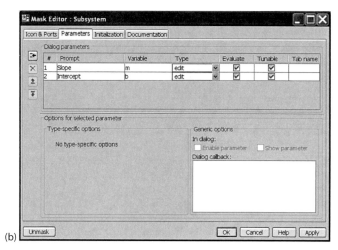

(b)

FIGURE 8.16   The final masked subsystem example.   (*Continued*)

lite for a 48-hour period. (Hint: consider the use of simulation-time-controlled switches.)

2. Repeat the simulation above for a *tangential* square impulse force. Show the 48-hour orbit for the satellite.

8.3 CREATING LIBRARIES

Now that we have a new parameterized subsystem, we may discover that it is of general use and should be made available to everyone. To make it available, we would like to add it to a *library*. Libraries in Simulink are just

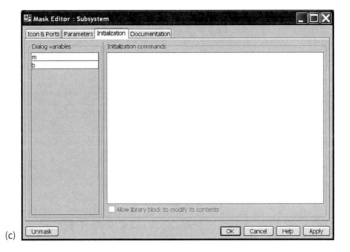

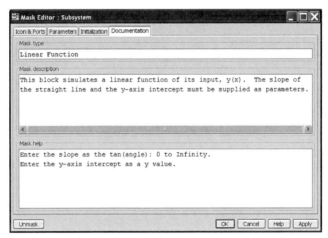

FIGURE 8.16 (Continued)   The final masked subsystem example.

model files that have been described to Simulink by an slblocks.m file and reside in a folder whose path has been registered with MATLAB.

If we want to add the block to a library, all we have to do is locate the library model file where we want the block and copy the block into the model, just as we copy library blocks into our model windows.

There are two approaches to this task. First, we could simply add it to an existing library in Simulink. For example, we could add it to the User-Defined Functions library. To do this, we open up our model in one window. Then we locate and open the Simulink model file, named `simulink.mdl`. It resides in folder `<MATLAB folder path>\tool-box\simulink\blocks` (for Release R2009b). (On the author's system, `<MATLAB folder path>` is `C:\Program Files\MATLAB\R2009b 255197`.) The situation at this point looks like Figure 8.17.

We see that the Simulink library contains a number of subsystems, which are the same as those appearing on the right in the Library Browser. If we double-click the `User-Defined Functions` subsystem, we will see Figure 8.18.

Now we can select our newly defined subsystem block and drag it into the `User-Defined Functions` window. This may require us to unlock the library. This can be done via the `Edit | Unlock Library` command or by response to a query box after we add it. Figure 8.19 shows the block added to `User-Defined Functions`. If we now save the Simulink library model, Figure 8.20 shows the final availability in the Simulink Library Browser.

It might be the case that we desire a separate new library that could be distributed to our users. In this situation, we use the second approach in which we create an entirely new library, add our block to it, and add the new library to the Simulink Library Browser. To accomplish this, we execute the `File | New | Library` command to create a model file that is marked as a library. Completion of this command yields the new-library model window shown in Figure 8.21.

We must now add the new block to the new empty library and save the library in a folder. Now we must inform MATLAB that it should store the folder path. We can do this with the addpath command at a prompt in the MATLAB

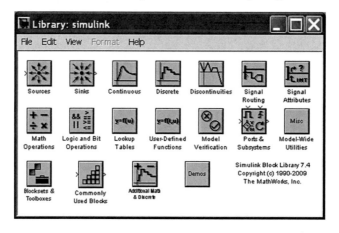

FIGURE 8.17 The Simulink library file opened.

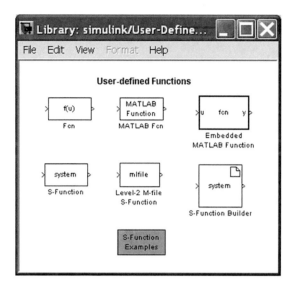

FIGURE 8.18   The User-Defined Functions subsystem opened.

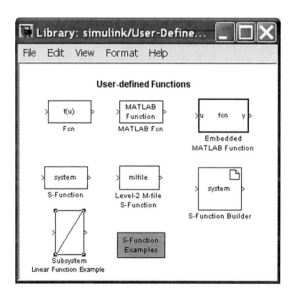

FIGURE 8.19   The User-Defined Functions subsystem modified.

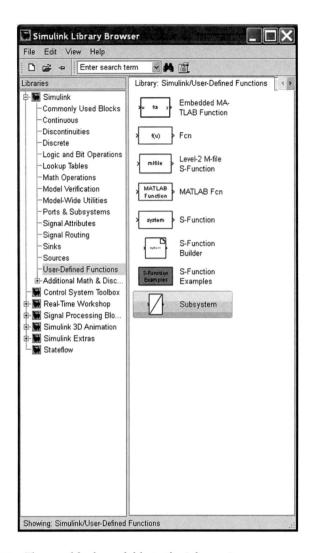

FIGURE 8.20 The new block available in the Library Browser.

Command window. The command in this example is addpath('C:\ Program Files\MATLAB\NewLibrary'). This command must be executed each time MATLAB is started, unless it has been added as part of the standard MATLAB initialization (see MATLAB documentation).

Next we must add the slblocks.m file to this same folder so that the Simulink browser can locate and process the library. The best way to do this is to copy an existing file. A good one to use in the Student system is the Stateflow® Library slblocks.m file shown in Figure 8.22.

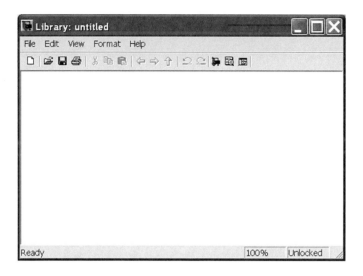

FIGURE 8.21   An empty library.

In Figure 8.22, we see the file and library names that we need to replace with our new file and library names. After we edit it, replacing the file and library names, we have Figure 8.23. Note that the changed program text is shown in boldface.

Restarting Simulink will reveal that a new library exists in the Simulink Library Browser with the name New Library, and it can be opened to reveal the new block available for use, as we see in Figure 8.24.

EXERCISE 8.3

A particle with charge Q and mass m is held at the origin of a coordinate system with an electric field E pointing along the z-axis and a magnetic field B pointing along the x-axis. If the particle is released at time 0, it will move in the y-z plane according to the model

$$\ddot{y}(t) = \omega\dot{z}(t) \quad \dot{y}(0) = 0 \quad y(0) = 0 \tag{8.7}$$

$$\ddot{z}(t) = \omega\left(\frac{E}{B} - \dot{y}(t)\right) \quad \dot{z}(0) = 0 \quad \dot{z}(0) = 0 \tag{8.8}$$

where $\omega = QB/m$ is called the cyclotron frequency (Griffiths 1999). Simulate the behavior of the particle when the ratio $E/B = 1.33$ in

```
function blkStruct = slblocks
%SLBLOCKS Defines the block library for a specific Toolbox or Blockset.

%   Copyright 1995-2004 The MathWorks, Inc.
%   $Revision: 1.16.2.2 $  $Date: 2004/04/15 01:01:51 $

% Name of the subsystem which will show up in the SIMULINK Blocksets
% and Toolboxes subsystem.
% Example:  blkStruct.Name = 'Signal Processing Blockset';

% The function that will be called when the user double-clicks on
% this icon.
% Example:  blkStruct.OpenFcn = 'dsplib';

% The argument to be set as the Mask Display for the subsystem.  You
% may comment this line out if no specific mask is desired.
% Example:  blkStruct.MaskDisplay = 'plot([0:2*pi],sin([0:2*pi]));';

blkStruct.MaskDisplay = 'disp(''SF'')';
blkStruct.OpenFcn = 'sflib';
blkStruct.Name = 'Stateflow';

if exist('sflib') == 4,
      Browser(1).Library = 'sflib';
   Browser(1).Name    = 'Stateflow';
   Browser(1).IsFlat  = 1;
      blkStruct.Browser = Browser;
end;

% End of slblocks
```

FIGURE 8.22   The standard `slblocks.m` file from the Stateflow® Library.

units of 10^6 m/s and $\omega = 5$ in units of 10^6 rad/s, giving $y(t)$ and $z(t)$ in units of meters. Display $y(t)$ and $z(t)$ individually in scopes and $y(t)$ versus $z(t)$ in an XY graph with $z(t)$ on the y-axis of the graph.

8.4 SUMMARY

We continued our study of second-order model simulations and looked at central-force systems. Orbital applications arise from central forces and are of both theoretical and practical importance to our planet. An important question in astronomical simulations is the choice of units so that numerical values remain within the best range for computation.

```
function blkStruct = slblocks
%SLBLOCKS Defines the block library for a specific Toolbox or Blockset.

%   Copyright 1995-2004 The MathWorks, Inc.
%   $Revision: 1.16.2.2 $  $Date: 2004/04/15 01:01:51 $

% Name of the subsystem which will show up in the SIMULINK Blocksets
% and Toolboxes subsystem.
% Example:  blkStruct.Name = 'Signal Processing Blockset';

% The function that will be called when the user double-clicks on
% this icon.
% Example:  blkStruct.OpenFcn = 'dsplib';

% The argument to be set as the Mask Display for the subsystem.  You
% may comment this line out if no specific mask is desired.
% Example:  blkStruct.MaskDisplay = 'plot([0:2*pi],sin([0:2*pi]));';

blkStruct.MaskDisplay = 'disp(''NL'')';
blkStruct.OpenFcn = 'NewLibrary';
blkStruct.Name = 'NewLibrary';

if exist('NewLibrary') == 4,
      Browser(1).Library = 'NewLibrary';
   Browser(1).Name    = 'New Library';
   Browser(1).IsFlat  = 1;
      blkStruct.Browser = Browser;
end;

% End of slblocks
```

FIGURE 8.23 The modified **slblocks.m** file to be placed in the New Library folder.

We investigated how we can make our own blocks by using masked subsystems, which allow the simulator to create parameterized blocks. These blocks can be included in existing libraries by inserting them into the file folders used by Simulink. We also examined how to make our own libraries to contain these customized blocks.

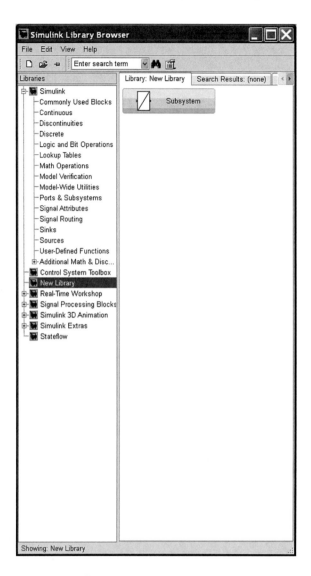

FIGURE 8.24 The modified Simulink Library Browser showing the New Library.

REFERENCES

Dym, C. 2004. *Principles of Mathematical Modeling*. New York: Elsevier Academic.

Garcia, A. 2000. *Numerical Methods for Physics*. Upper Saddle River, NJ: Prentice Hall.

Gould, H., and J. Tobochnik. 1996. *An Introduction to Computer Simulation Methods: Applications to Physical Systems*. New York: Addison Wesley.

Griffiths, D. J. 1999. *Introduction to Electrodynamics*. Upper Saddle River, NJ: Prentice Hall.

Haberman, R. 1977. *Mathematical Models: Mechanical Vibrations, Population Dynamics, and Traffic Flow*. Englewood Cliffs, NJ: Prentice Hall.

Higham D. J., and N. Higham. 2005. *MATLAB Guide*. Philadelphia: Society for Industrial and Applied Mathematics.

Knight, R. D., B. Jones, and S. Field. 2007. *College Physics: A Strategic Approach*. New York: Pearson Addison Wesley.

CHAPTER 9

Higher-Order Models and Variable-Step Solvers

IT IS UNCOMMON TO encounter systems described by differential equations higher than second order. But such systems do occur, so we now consider how we can create simulations of such systems. Using the techniques we have seen for second-order systems, we find that systems whose dynamical models are higher order than second order can be simulated just as easily as first- and second-order equations.

9.1 DIRECT SIMULATION BY MULTIPLE INTEGRATIONS

Earlier in this book, we examined the direct simulation of second-order equations by multiple integrations, and it is equally possible to use the same approach for differential equations with an order higher than two.

In Chapters 4 and 6, our approach was to rewrite the dynamical equation by isolating the highest-order derivative term on the left-hand side and putting the other terms on the right-hand side. This technique can be used for higher-order models simply by increasing the number of integrators to the order of the highest-order term and arranging each so that its output feeds the input to the next in line. The use of multiple integrators makes the problem similar to the lower-order models. One integrator for each order of the model is included so that higher-order models will simply have more integrators and more initial conditions. We can demonstrate the use of this same technique on a third-order equation.

9.1.1 An Automobile Suspension Model

As an example of a higher-order model, consider a simple model of an automobile suspension system described by Dorf and Bishop (2008). This system consists of a wheel attached to an axle, and the axle is attached to the automobile body by a spring and a shock absorber. The shock absorber and the spring both provide spring actions with spring constants k_1 and k_2. The shock absorber also provides a linear vibrational damping action proportional to the vertical speed, $\dot{y}(t)$, with a damping constant f. The mass supported by the wheel is M. We would like to model that action of the body as the wheel passes over a bump in the road.

An analysis of the forces on this system yields the following third-order equation for the vertical displacement of the body, $y(t)$:

$$\dddot{y}(t) = -\left(\frac{k_1 + k_2}{f}\right)\ddot{y}(t) - \frac{k_2}{M}\dot{y}(t) - \frac{k_2 k_1}{Mf}y(t) \tag{9.1}$$

The model for this system is shown in Figure 9.1.

As an example, we use $k_1 = 7500$ and $f = 1000$ for the shock absorber, $k_2 = 150{,}000$ for the spring, and $M = 500$ for the mass supported by the wheel. We examine the action after going over a drop of 0.25 feet at $t = 0$. The result is shown in Figure 9.2.

9.1.2 The Terminator Block from the Sinks Library

In Model_9_1_1 for the automobile suspension, we have included a subsystem for all the constants and their combinations, which considerably simplifies the visual appearance of the model. Expanding this subsystem, we find the structure in Figure 9.3.

Notice that we have organized the constants in this model as a series of bus-like input lines that can be "tapped" to provide the inputs for the various combinations they are involved in. The lines propagate across the entire bottom to make it clear where the constant signals are going. This structure is a common one in electrical design diagrams and is very useful here. But when Simulink® is asked to execute the simulation, it will detect that the lines are unconnected and give a warning. To eliminate this warning, we can use the Terminator block from the Sinks library. This block

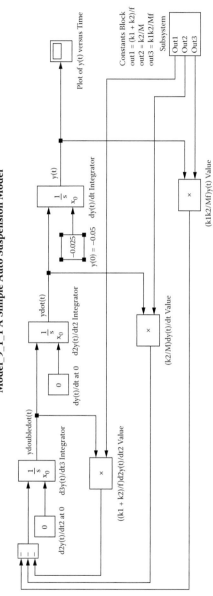

FIGURE 9.1   Simulation model for an automobile suspension.

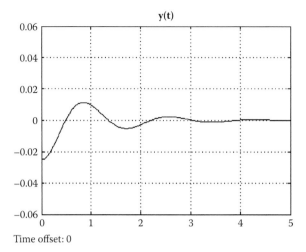

FIGURE 9.2 The response of an automobile suspension passing over a 0.25-ft bump. The *y*-axis shown the displacement of the supported mass in feet, with the *x*-axis showing the simulation time in seconds.

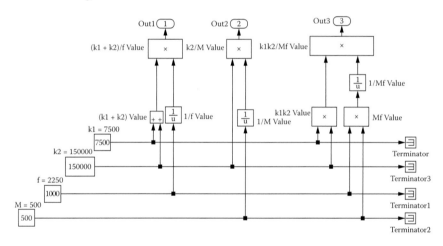

FIGURE 9.3 Constant subsystem for the automobile suspension.

acts like an electrical termination for electrical signal lines and "absorbs" the signal into the block. Simulink now finds that the connections are all complete and runs the model without complaint. The Terminator blocks have no parameters and can only terminate one signal, so their use in a diagram is very simple.

9.1.3 Transformation into a First-Order Equation System

A second way to produce a simulation structure is to use a transformation of each nth-order differential equation into n first-order equations. Then, the same first-order structures we have been using earlier can be used to make the nth-order simulation.

It is well known that an nth-order differential equation can be transformed into n first-order differential equations by using a set of variable transformations. As an example, we begin with a third-order differential equation

$$\dddot{y}(t) + c(t)\ddot{y}(t) + d(t)\dot{y}(t) + e(t)y(t) = f(y(t), t) \tag{9.2}$$

with initial values $y(0) = g$, $\dot{y}(0) = h$, and $\ddot{y}(0) = k$. We now define a set of transformation equations by defining three new variables: $x_1(t) = y(t)$, $x_2(t) = \dot{y}(t)$, and $x_3(t) = \ddot{y}(t)$ (Severance 2001). These three variables must be the solution to the coupled first-order equations

$$\dot{x}_1(t) = x_2(t) \tag{9.3}$$

$$\dot{x}2(t) = x_3(t) \tag{9.4}$$

$$\dot{x}_3(t) = -c(t)x_3(t) - d(t)x_2(t) - e(t)x_1(t) + f(t) \tag{9.5}$$

with the initial values $x_1(0) = g$, $x_2(0) = h$, and $x_3(0) = k$. This set of coupled first-order equations produces the same function for $x_1(t)$ that the third-order Equation (9.2) produces for $y(t)$. So a simulation of the coupled first-order equations will be identical to a simulation of the third-order equation.

We can immediately write the simulation for these coupled, first-order equations in the Simulink model in Figure 9.4. When we examine this structure, we realize that it is the same structure as the one we would have created for the direct simulation of the third-order Equation (9.2). So the coupled first-order system simulation produces exactly the same result as the direct third-order simulation.

Structure for Simulating Coupled First-Order Differential Equations

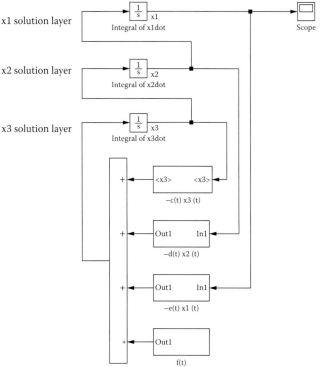

FIGURE 9.4   Simulation of coupled, first-order differential equations.

9.2 PRODUCING FUNCTION FORMS FOR SIMULATION RESULTS

Up to now, we have been content with displaying simulation results in numeric form and graphical visualizations. But it is often important to produce a functional form for the simulation results, since the function may give important information about the system dynamics. In this section, we examine how we can use a MATLAB® curve-fitting function to produce a function for a simulation output.

Let's use Model_7_4_3, the model of the object falling in a uniform gravitational field with an initial velocity, as our example. This model produces the results shown in Figure 9.5.

Suppose we would like to produce a function for the trajectory that the object follows. We can run the simulation and use the To Workspace block to send the $x(t)$ and $y(t)$ values to the MATLAB workspace. If we use the MATLAB variables x and y to save these values, we can recover them

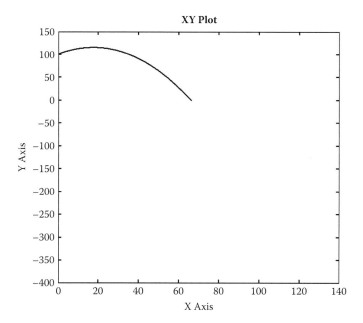

FIGURE 9.5   Results of the realistic simulation of Model_7_4_3.

```
>> p=polyfit(x.signals.values(1:663),y.signals.values(1:663),2)

p =

   -0.0490    1.7321    99.9998

>>
```

FIGURE 9.6   The execution of the polyfit function and its results.

using the names x.signals.values and y.signals.values. The MATLAB function we can use to fit these two sets of data is the polyfit function. This function takes three arguments: a set of independent values x_i, a set of function values y_i for the independent values, and the degree of a polynomial to fit to the data. The polyfit function finds the coefficients of the polynomial that gives the closest fit by minimizing the squared errors for the fit. If y_i differs from $p(x_i)$ by an error ε_i, the fitting algorithm calculates the sum of the squares of the errors ε_i at each point and finds the coefficients that minimize this sum.

Figure 9.6 shows the MATLAB function execution with the returned coefficients for our example. We can see that the degree 2 polynomial that fits the data best is

$$y = -0.0490x^2 + 1.7334x + 99.9888 \tag{9.6}$$

If we plot the polynomial found by the `polyfit` function and the simulation results in Figure 9.7, we find an excellent fit, as we expect, since the trajectory of a falling object should be a parabola.

The coefficients found by the algorithm approximate the theoretical values almost exactly, since the theoretical function is

$$y = \frac{1}{2}\frac{g}{v_0^2 \cos^2 \theta_0}x^2 + \tan\theta_0 x + h \tag{9.7}$$

and substituting the values $v_0 = 20$ *m/s*, $\theta_0 = 60°$, and $g = 9.8$ *m/s²* into the theoretical expressions gives

$$\frac{1}{2}\frac{g}{v_0^2 \cos^2 \theta_0} = -0.049, \quad \tan\theta_0 = 1.7321, \quad h = 100 \tag{9.8}$$

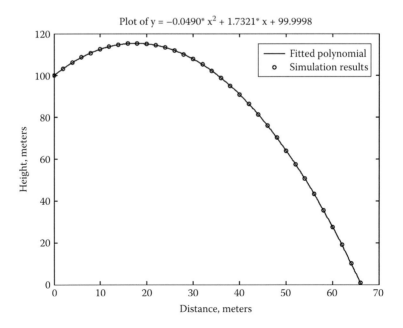

FIGURE 9.7 The plot of the `polyfit` function and the simulation results.

Fitting data to functional forms is a large and complicated issue in mathematics, so we only briefly introduce this technique, and the student is encouraged to do additional reading in this area before using the technique routinely.

EXERCISE 9.1

Find a polynomial fit to the $y(t)$ component of the simulation results produced by Model_7_4_3 and verify that the coefficients are close to the theoretical values of the equation

$$y(t) = -\frac{1}{2}gt^2 + v_0 \sin\theta_0 t + h \qquad (9.9)$$

using the values given above for the constants.

9.3 VARIABLE-STEP SOLVERS

When fixed-step solvers are used, the step size taken by the solver must be set to the smallest step size required in the problem, even though there may be times in the problem solution when the solution curve is changing very slowly. If the solution is changing slowly, a larger step size would be more desirable, since it would lead to a faster solution with less work. But the fixed size of the step in the fixed-step methods prevents us from doing this.

The solution to this dilemma is to devise methods that do not require fixed step sizes but change the step size according to the problem's conditions at that point. These *adaptive* methods are employed in Simulink's *variable-step solvers*. The basic solution methods are essentially the same as those used in the fixed-step solvers, but extra computations are made at the beginning of a step to decide what the size of the step should be. Clearly, additional computations in each step will slow the step-generation time down, but if we can reduce the number of steps, there may be an overall saving in time.

In addition to the savings in time, a very important issue for variable-step methods is the question of truncation error. As we saw in Chapter 5, the truncation error we acquire in a simulation is proportional to a power of the step size. For a fourth-order *Runge–Kutta* solver, the truncation error will accumulate as the fifth power of the step size, $\varepsilon \propto \tau^5$, where ε is the global truncation error and τ is the step size. Because of this, variable-step methods must be very careful not to use too large a step size in parts of the simulation when the solution values are changing rapidly. To ensure that the simulation maintains accuracy, the algorithms must maintain accuracy during step size change.

9.3.1 Example Variable-Step-Size Algorithm

As an example of an algorithm for determining the step size appropriate for the required accuracy, we could use the following procedure. We assume that we have a step size, τ_{last}, from the last step we made. We want to determine a new step, τ_{next}, for the next step, which can be different from the last step.

We can write an equation for the new step size as

$$\tau_{next} = M_1 \tau_{last} \tag{9.10}$$

where M_1 is a multiplier that will increase or decrease the size of the last step.

We can determine the value of M_1 by using the difference in truncation errors as a guide to the value. First, we use our numerical integration formula to compute the next value, $y_{\tau_{last}}$, using τ_{last} as the step size. Then we recompute the value, $y_{\tau_{last}}$, using $\frac{1}{2}\tau_{last}$ as the step size, making two computations to span the step τ_{last}. Then we form the difference of the truncation errors of the two step sizes using

$$\varepsilon_{last} = \left| y_{\tau_{last}} - y_{\frac{1}{2}\tau_{last}} \right| \tag{9.11}$$

If the error desired by the user of the solver is $\varepsilon_{desired}$, we can form the ratio of the two errors and use it to compute M_1 using

$$M_1 = f\left(\left| \frac{\varepsilon_{desired}}{\varepsilon_{last}} \right| \right) \tag{9.12}$$

and then use M_1 to find τ_{next} using Equation (9.10).

9.3.2 Example Variable-Step, Fourth-Order *Runge–Kutta* Solver

Let's consider how we can use this approach in making a variable-step *Runge–Kutta* solver. Since the global truncation error for a fourth-order *Runge–Kutta* method varies as the step size to the fifth power, we take the function in Equation (9.12) to be the fifth root of the ratio of the errors.

$$M_1 = \left(\left| \frac{\varepsilon_{desired}}{\varepsilon_{last}} \right| \right)^{\frac{1}{5}} \tag{9.13}$$

We need to be careful not to increase the step size too much nor decrease it too much, so we set upper and lower limits on the change by using limiting factors M_{upper} and M_{lower}. Putting all this together, we have the equations

$$\tau_{next} = M_{upper} \tau_{last} \quad \text{when} \ M_{upper} \tau_{last} < M_1 \tau_{last} \tag{9.14}$$

$$\tau_{next} = M_1 \tau_{last} \quad \text{when} \ M_{lower} \tau_{last} \le M_1 \tau_{last} \le M_{upper} \tau_{last} \tag{9.15}$$

$$\tau_{next} = M_{lower} \tau_{last} \quad \text{when} \ M_1 \tau_{last} < M_{lower} \tau_{last} \tag{9.16}$$

If we examine the parameters for Simulink's `ode45` solver, we see that the simulator enters values like these as the parameters `Max step size`, `Min step size`, and `relative tolerance` into the solver (see Figure 9.8).

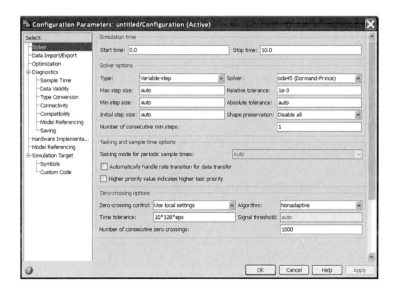

FIGURE 9.8   `ode45` solver parameters.

9.4 VARIABLE-STEP SOLVERS IN SIMULINK

Simulink provides two main variable-step solvers for the simulator. They are `ode45` (Dormand–Prince) and `ode23` (Bogacki–Shampine). Both of these solvers employ sophisticated interpolation methods for determining the next step size, which involve both desired relative and absolute errors. The simulator can allow Simulink to make decisions for the absolute error. The Simulink documentation notes that

- `ode45` is based on an explicit *Runge–Kutta* (4,5) formula, the Dormand–Prince pair. It is a *one-step* solver—in computing y(tn), it needs only the solution at the immediately preceding time point, y(tn−1). In general, `ode45` is the best function to apply as a "first try" for most problems.

- `ode23` is an implementation of an explicit *Runge–Kutta* (2,3) pair of Bogacki and Shampine. It may be more efficient than `ode45` at crude tolerances and in the presence of moderate stiffness. Like `ode45`, `ode23` is a one-step solver.

It is beyond the scope of this book to consider the problem of numerical solution to stiff differential equations, so we will simply note that some differential equations have solutions that involve multiple rates of change, some slow and some fast. These kinds of differential equations are called *stiff* equations because the problems in which they were first defined were the solutions of the dynamics of stiff vibrating bars. The nonstiff solver has considerable difficulty in solving these kinds of equations, so special solvers must be used for these dynamical systems.

9.5 SUMMARY

Models with equations higher than second order are infrequent. The direct simulation of a higher-order differential equation is achieved by a simulation layer with a sequence of integrators. The number of integrators must be equal to the order of the highest-order derivative. A higher-order differential equation can be transformed into a coupled, first-order equation system through a change of variables. The system of equations can be simulated as a set of interconnected layers, one for each first-order equation. We studied how variable-step solvers improve the performance of our simulations. Simulink offers a set of these solvers for use in appropriate situations. We also examined techniques for solving differential equations of higher order than two.

We briefly introduced the idea of fitting simulation results to a functional form. We used the polyfit function in MATLAB to find a polynomial fit to a model's outputs.

Variable-step solvers analyze the rate of change of the solution to decide on the next step size. The overhead of step-size analysis can be lower than the overhead of maintaining a fixed step-size, leading to an overall gain in efficiency.

REFERENCES AND ADDITIONAL READING

Dorf, R. C., and R. Bishop. 2008. *Modern Control Systems*. Upper Saddle River, NJ: Prentice Hall.

Klee, H. 2007. *Simulation of Dynamic Systems with MATLAB and Simulink*. Boca Raton, FL: CRC Press.

Severance, F. 2001. *System Modeling and Simulation: An Introduction*. New York: John Wiley.

The MathWorks, Inc. 2007. http://www.mathworks.com.

Advanced Topics

Transforming Ordinary Differential Equations, Simulation of Chaotic Dynamics, and Simulation of Partial Differential Equations

10.1 TRANSFORMING ORDINARY DIFFERENTIAL EQUATIONS

We have been studying the simulation of dynamic systems from a common approach taken in fields of science and mathematics. This is an analysis of the physical laws that determine a system's behavior leading to a set of dynamical equations, whose solution tells us how the system evolves in time. Once we have this system of equations, we analyze their solutions using mathematically developed techniques by solving the equations in the physical time continuum. But the solution of dynamical equations in time and space, that is, in a space where time is the independent variable, is often very difficult to achieve. As a result, another approach has been developed that depends on *transforming* the equations into a different independent-variable space. Once the equations are expressed in the different space, assuming that we have been clever about choosing the new space, they are easier to solve in the new space than in time and space.

Engineers, in particular, have adopted this approach on a broad basis, and *control theory* (Klee 2007) uses this approach in designing systems that do useful work under control. Simulations of other kinds of systems may also use this approach, so it is worth our effort to learn something about it.

We simply introduce this important subject here so that students will have a nodding acquaintance with transforms in their simulation study.

Let's assume that we have produced an nth-order ordinary differential equation that describes the dynamics of a system in time, and that it has the form

$$c_n \frac{d^n f(t)}{dt^n} + c_{n-1} \frac{d^{n-1} f(t)}{dt^{n-1}} + \ldots + c_1 \frac{df(t)}{dt} + c_0 f(t) = g(t) \qquad (10.1)$$

We have been viewing this equation as a set of terms that specify how the value of a function and its derivatives are related along the time axis. In the case of Equation (10.1), the sum of the function value (and the associated multiplier) and the value of its derivatives up to order n must produce the function $g(t)$ at each point on the time axis.

An alternative way to view this equation is that it is the transformation of an *input function*, $g(t)$, into an *output function*, $f(t)$. If we took this point of view, we would write Equation (10.1) in the form

$$f(t) = \frac{1}{c_n \dfrac{d^n}{dt^n} + c_{n-1} \dfrac{d^{n-1}}{dt^{n-1}} + \ldots + c_1 \dfrac{d}{dt} + c_0} g(t) \qquad (10.2)$$

This looks odd when we first see it because the idea of an inverse for a derivative operation is not an easily understood concept. But let's think of the reciprocal of the derivative operators as a function that operates on the function $g(t)$ to produce the function $f(t)$. We won't concern ourselves in this text about the conditions that must be satisfied for such an interpretation to be meaningful, but the works listed in the References and Additional Reading at the end of this chapter can be consulted for more information.

The first term on the right-hand side looks like an operator acting on an input to produce an output or, alternatively, transferring an input to the output. Because of this analogy, the reciprocal differential operator is called the *transfer function* of the system. This input–output view is particularly appealing to engineers, who are usually engaged in trying to design a system to perform such an action, and Equation (10.2) is often abbreviated as

$$f(t) = H(t)g(t) \tag{10.3}$$

If we are given a system like this, we want to know whether a transformation of the system into another space would make it easier to find the output function, $f(t)$. That is, if we transformed both sides of Equation (10.3) from a *t-space* to an *s-space*, then we would have

$$Transform_{t \to s}\, [f(t)] = Transform_{t \to s}\, [H(t)g(t)] \tag{10.4}$$

and we would like to know whether the transformed equation's solution is easier to find in the new *s-space*.

Suppose that we restrict our systems to those for which the transform of a product is identical to the product of the transforms. Then we can assert

$$Transform_{t \to s}\, [f(t)] = Transform_{t \to s}\, [H(t)] = Transform_{t \to s}\, [g(t)] \tag{10.5}$$

and we can write Equation (10.4) as

$$f^{\star}(s) = H(s)g^{\star}(s) \tag{10.6}$$

where

$$f^{\star}(s) = Transform_{t \to s}[f(t)] \tag{10.7}$$

$$g^{\star}(s) = Transform_{t \to s}[g(t)] \tag{10.8}$$

$$gH(s) = Transform_{t \to s}[H(t)] \tag{10.9}$$

We then need to solve the transformed Equation (10.6) for the solution $f^{\star}(s)$.

Now scientists and mathematicians have found that, if the system possesses certain properties, it is indeed possible to transform a problem, and the resulting equation is easier to analyze. But, if we transform the variable into a different variable, how do we relate the solution we find, $f^{\star}(s)$, which is in a different space, to a function in physical time and space? What we must do is perform the *inverse transformation* to return the solution to physical time, as seen in the following equation:

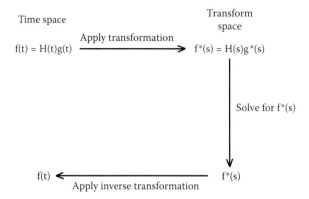

FIGURE 10.1 The transformation model. Asterisks indicate the transformed functions.

$$f(t) = Transform_{s \to t}^{-1}[f*(s)] \tag{10.10}$$

A graphical portrayal of the method is shown in Figure 10.1.

This is actually not so unusual in the science world, since, even in elementary discussions, a problem in Cartesian coordinates may be transformed into polar coordinates because it is easier to solve in those coordinates. Then, the resulting solution in polar coordinates can be transformed back into Cartesian coordinates for its final form. That is exactly the process described here.

Returning to the transformation of an ordinary differential equation in time into a different space, we restrict our systems to those that are *linear* and *homogeneous* because the properties of these kinds of systems are exactly those we need for our transformation model.

A linear system has the property: if an input $g_1(s)$ produces the output $f_1(s)$, and a different input $g_2(s)$ produces a different output $f_2(s)$, then the combined input $g_1(s) + g_2(s)$ will produce the output $f_1(s) + f_2(s)$.

This property implies that we can write

$$f_1(t) + f_2(t) = H(t)(g_1(t) + g_2(t)) \tag{10.11}$$

This is called *the principle of superposition*. It means that the output of any input that can be represented as a sum of basic inputs can be produced by finding the outputs of the basic inputs individually and summing them.

A homogeneous system has the property: if the input g(t) produces the output f(t), then the output produced by multiplying the original input by a constant c, will be the original output multiplied by the same constant c.

This property implies that we can write

$$cf(t) = H(t)\,(cg(t)) \tag{10.12}$$

The important thing about these systems is that we can decompose the input into a sum of basic inputs, operate on each basic input independently to produce an output, and then sum the resulting outputs to form the final solution.

There are a number of transformation types that have been used by analysts in solving problems. The most well known are the Laplace and Fourier transforms. The Fourier transform is used to analyze the output of system execution, rather than the dynamical development of a physical system. The Laplace transform, on the other hand, is an important means of simulating the dynamics of systems. In this book, we focus on the Laplace transform.

10.1.1 The Laplace Transform

This transform is the most general of the two and is commonly used in engineering applications because it is particularly useful in analyzing the time behavior of systems into transient and steady-state parts. It is defined by

$$f^*(s) = \int_0^\infty f(t)e^{-st}\,dt \tag{10.13}$$

where s is an independent complex variable. An interpretation of this transform is that it maps the function with a particular weight, e^{-st}, along the entire time axis into a single point, s, in the *complex plane*, where each s is written $s = x + iy$.

The complex plane is a way of graphically representing complex numbers and their functions in the same way that real numbers and their functions are represented. But since complex numbers have two parts, the real part x and the complex part y, they must be represented by a two-dimensional plane instead of a one-dimensional line as the real numbers are. We view an (x,y) point in the complex plane as representing the complex number s = $x + iy$ and a function value $f(s)$ as lying on a surface above the complex plane. This is the two-dimensional analog to the one-dimensional presentation of the real numbers and their functions. Just as the integration of a real function is along the one-dimensional axis for real numbers, the integration of a complex function is over an area of the complex plane for complex numbers. Scientists and engineers have developed a powerful set of mathematical tools to analyze complex functions, and the study of these techniques is the subject of advanced courses, so we just introduce their use in this brief survey of simulation using transform techniques.

The inverse Laplace transform is

$$f(t) = \frac{1}{2\pi i} \int_{c-i\infty}^{c+i\infty} f*(s)e^{st}\,ds \tag{10.14}$$

This maps the complex function $f(s)$ over the entire complex plane with a weight e^{st} back into a single point t on the time axis. As noted previously, integration in the complex plane is a mathematically sophisticated affair, but the good news about this transformation is that we rarely have to perform the integrals, since engineers over the years have built up large tables of transform pairs that we can use to look up transformed functions. As an example, Table 10.1 shows the pairs for some common functions.

TABLE 10.1 Some Laplace Transform Pairs

$f(t)$	$f(s)$
1	$1/s$
t	$1/s^2$
$t^{n-1}/(n-1)!$	$1/s^n$
e^{-at}	$1/s+a$
$\cos(at)$	s/s^2+a^2

From our perspective, we are most interested in how derivatives with respect to time are transformed, since the differential operator, $H(t)$, will have to be transformed as well. The Laplace transformation for derivatives and integrals is given by

$$\text{Laplace transform}\left[\frac{df(t)}{dt}\right] = sf^*(s) - f(t)|_{t=0} \tag{10.15}$$

$$\text{Laplace transform}\left[\frac{d^2f(t)}{dt^2}\right] = s^2 f^*(s) - sf(t)|_{t=0} - \frac{df(t)}{dt}|_{t=0} \tag{10.16}$$

$$\text{Laplace transform}\left[\int_0^t f(t)dt\right] = \frac{1}{s}f^*(s) \tag{10.17}$$

10.1.2 A Laplace Transform Example

As an example of the use of the Laplace transform, let's solve by hand the following differential equation:

$$\frac{d^2 f(t)}{dt^2} - 6\frac{df(t)}{dt} + 9f(t) = 2t \tag{10.18}$$

with the initial conditions

$$f(0) = 0 \quad \text{and} \quad \frac{df(t)}{dt}|_{t=0} = 0 \tag{10.19}$$

We transform both sides of Equation (10.19) using Equations (10.15), (10.16), and Table 10.1 and find

$$s^2 f^*(s) - sf(t)|_{t=0} - \frac{df(t)}{dt}|_{t=0} - 6\left[sf^*(s) - f(t)|_{t=0}\right] + 9f^*(s) = \frac{2}{s^2} \tag{10.20}$$

Substituting the initial conditions, we have

$$s^2 f^*(s) - 6sf^*(s) + 9f^*(s) = \frac{2}{s^2} \qquad (10.21)$$

Solving for $f(s)$ gives

$$f^*(s) = \frac{2}{s^2}\frac{1}{s^2 - 6s + 9} \qquad (10.22)$$

which we can write as

$$f^*(s) = \frac{2}{s^2(s-3)^2} \qquad (10.23)$$

At this point, we would like to invert both sides to find the solution $f(t)$. To do this, we would like to look up the inverse transform of the right-hand side. Examining tables of transforms, we find that terms like

$$\frac{c}{s^2} \quad \text{and} \quad \frac{c}{s-d}$$

have been worked out and appear in the published table, but that complicated expressions like the right-hand side of Equation (10.23) do not. This requires us to simplify the right-hand side, and we can use the method of partial fractions to do this.

The method of partial fractions is based on the recognition of a ratio of polynomials in s, such as

$$\frac{as^m + bs^{m-1} + \ldots + g}{cs^n + ds^{n-1} + \ldots + h}$$

where $m < n$, can be reduced to a sum of simpler ratios by finding the roots of the polynomial in the denominator. The roots of the denominator $cs^n + ds^{n-1} + \ldots + h$ can be found by solving the equation

$$cs^n + ds^{n-1} + \cdots + h = 0 \qquad (10.24)$$

and finding a product expression like

$$(s-r_1)(s-r_2)\cdots(s-r_n)=0 \tag{10.25}$$

Then we can write the original polynomial ratio as

$$\frac{as^m + bs^{m-1} + \cdots + g}{(s-r_1)(s-r_2)\cdots(s-r_n)} \tag{10.26}$$

Now, by choosing an appropriate set of constants c_n, we can rewrite Equation (10.26) as

$$\frac{c_1}{s-r_1} + \frac{c_2}{s-r_2} + \cdots + \frac{c_n}{s-r_n} \tag{10.27}$$

This is called a partial-fraction expansion of the original ratio of polynomials. We have glossed over a number of important issues in this method that arise from the kinds of roots for the denominator polynomial by considering only the case where the roots are single, real, and distinct. The student can find a complete analysis in Klee (2007).

Returning to our original goal of finding the inverted solution for Equation (10.23), if the right-hand side is expanded in a partial-fraction expansion, we find

$$f^*(s) = \frac{\frac{2}{9}}{s^2} + \frac{\frac{4}{27}}{s} + \frac{\frac{2}{9}}{(s-3)^2} + \frac{-\frac{4}{27}}{s-3} \tag{10.28}$$

Now we can use a table to look up the inverse transform of each term separately to find the solution in time

$$f(t) = \frac{2}{9}t + \frac{4}{27} + \frac{2}{9}te^{3t} - \frac{4}{27}e^{3t} \tag{10.29}$$

We can check our exact solution with the simulation resulting from the multiple-integration method that we have been using by constructing the Simulink® model shown in Figure 10.2. The results for the time interval [0.0,2.0] are shown in Figure 10.3.

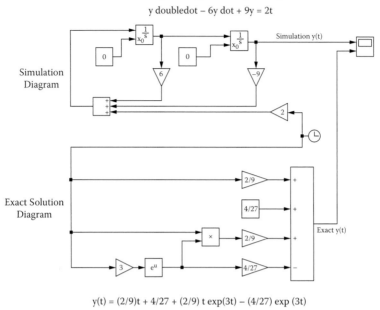

Model_10_1 Example Using the Multiple Integration Method

y doubledot − 6y dot + 9y = 2t

$$y(t) = (2/9)t + 4/27 + (2/9)\, t \exp(3t) - (4/27) \exp(3t)$$

FIGURE 10.2   Model_10_1_1 comparing the multiple integration simulation to the exact solution.

10.1.3 Simulation by Nested Transforms

Our interest is not in solving these equations by hand, but in constructing simulations using the Laplace transform. If we want to exert some manual effort to transform the dynamical equations before beginning the simulation, then we can construct simulation diagrams directly from the transformed equations. We can see how to do this by simulating the transformed equation of the previous example using an approach of Matko, Zupancic, and Karba (1992) that we call the *nested structure*. But first we generalize Equation (10.18) to

$$\frac{d^2 f(t)}{dt^2} - 6\frac{df(t)}{dt} + 9f(t) = g(t) \tag{10.30}$$

Applying the Laplace transform to Equation (10.30), we find

$$(s^2 - 6s + 9)f^*(s) - g^*(s) = 0 \tag{10.31}$$

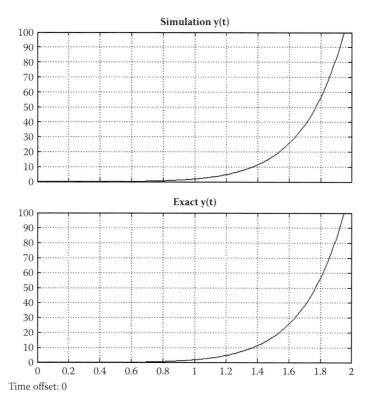

FIGURE 10.3 Results for Model_10_1_1 multiple integration method simulation. Values on the x-axis and y-axis are in arbitrary units.

Grouping terms of equivalent powers of s together gives

$$s^2\big(f^*(s)\big)+s\big(-6f^*(s)\big)+\big(9f^*(s)-g^*(s)\big)=0 \tag{10.32}$$

Next we take all the terms but the term that is highest order in s to the right-hand side.

$$s^2\big(f^*(s)\big)=-s\big(-6f^*(s)\big)-\big(9f^*(s)-g^*(s)\big) \tag{10.33}$$

We divide both sides by the highest-order power of s.

$$f^*(s)=\frac{1}{s}\big(6f^*(s)\big)+\frac{1}{s^2}\big(-9f^*(s)+g^*(s)\big) \tag{10.34}$$

Then, we regroup the terms on the right-hand side by taking out $1/s$ to get a nested form.

$$f^*(s) = \frac{1}{s}\left[6f^*(s) + \frac{1}{s}\left[g^*(s) - 9f^*(s)\right]\right] \tag{10.35}$$

Finally, we assign intermediate variables r_0, r_1 to the terms in the expression so that we can simplify it.

$$r_0(s) = \frac{1}{s}\left(g^*(s) - 9f^*(s)\right) \tag{10.36}$$

$$r_1(s) = \frac{1}{s}\left(6f^*(s) + r_0(s)\right) \tag{10.37}$$

$$f^*(s) = r_1(s) \tag{10.38}$$

We now have a form that can be implemented in a very regular structure. The simulation diagram is formed by replacing each term $1/s[...]$ by an Integrator block with the bracket expression as its input, as seen in Figure 10.4.

At this point, we can see why Simulink labels an Integrator block with the identifier $1/s$. It simulates the role of the $1/s$ term in a Laplace-transformed equation. The simulation and its results are shown in Figures 10.5 and 10.6.

Note that the nested structure is easily extended to higher-order transfer functions. Consider the transfer function

$$H(s) = \frac{as^3 + bs^2 + cs + d}{s^3 + es^2 + fs + g} \tag{10.39}$$

It is easy to write the nested structure by repeating the previously described method of defining intermediate variables. If we do, the simulation has a structure clearly exhibiting the regularity of this method, as seen in Figure 10.7.

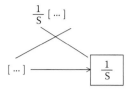

FIGURE 10.4   Mapping nested method terms into Simulink blocks.

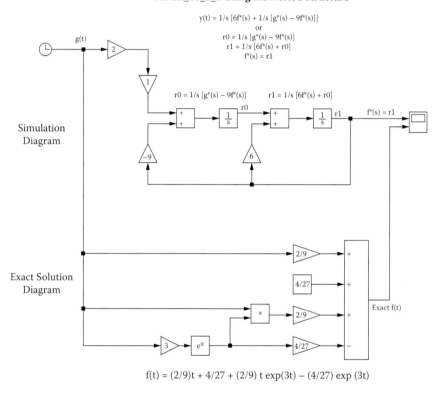

FIGURE 10.5   Model_10_2_1 for the nested structure simulation of a transfer function.

10.1.4 Simulation by Partitioned Transforms

Matko, Zupancic, and Karba (1992) give another method for implementing a simulation from the transformed dynamic equation that leads to a partitioned structure. In this method, we start with Equation (10.31) and rewrite it as

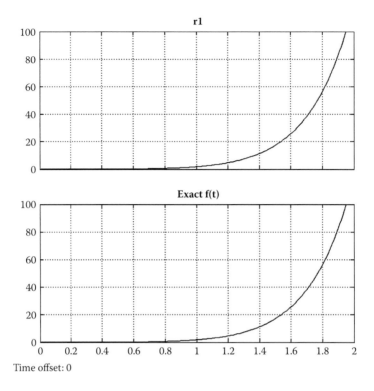

FIGURE 10.6 The nested structure simulation results for Model_10_2_1. Values on the x-axis and y-axis are in arbitrary units.

Model_10_3_1 Simulation (Nested Structure) of Transfer Function from Equation 10.35

$$a = b = c = d = 0.5; e = f = g = 1$$

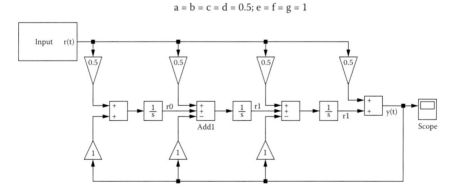

FIGURE 10.7 Model_10_3_1 for the nested structure for the transfer function of Equation (10.39).

$$\frac{f^*(s)}{g^*(s)} = \frac{1}{s^2 - 6s + 9} \tag{10.40}$$

Next, we introduce a new intermediate function, $d^*(s)$, by the equation

$$\frac{f^*(s)}{d^*(s)} \frac{d^*(s)}{g^*(s)} = \frac{1}{s^2 - 6s + 9} \tag{10.41}$$

Now we write two separate equations by equating the first ratio on the left-hand side to the numerator of the right-hand side, and the second ratio to the denominator.

$$\frac{f^*(s)}{d^*(s)} = 1 \tag{10.42}$$

$$\frac{d^*(s)}{g^*(s)} = \frac{1}{s^2 - 6s + 9} \tag{10.43}$$

We rewrite the second equation, Equation (10.43), into the following equation that can be simulated with a nested structure:

$$s^2 d^*(s) = g^*(s) + 6s d^*(s) - 9d^*(s) \tag{10.44}$$

With Equation (10.44), we can achieve an initial simulation diagram like that shown in Figure 10.8.

Next, we insert the first equation into the simulation diagram to connect the output $d^*(s)$ to the computation of $f^*(s)$.

$$f^*(s) = 1 \cdot d^*(s) \tag{10.45}$$

and finish the partitioned structure in Figure 10.9.

If we compare the output of this model with that of Model_10_2_1, we find that they are identical, as we expect. These structures are equivalent, and either approach can be used according to the preference of the simulator.

Model_10_4_1 Using the Partitioned Structure

$$(s^2 - 6s + 9)y(t) = g(t)$$

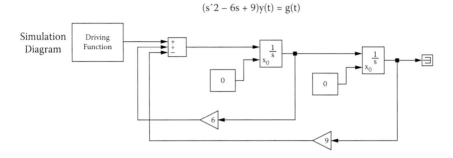

FIGURE 10.8 The initial Model_10_4_1 of the partitioned structure.

Model_10_4_2 Using the Partitioned Structure

$$(s^2 - 6s + 9)y(t) = g(t)$$

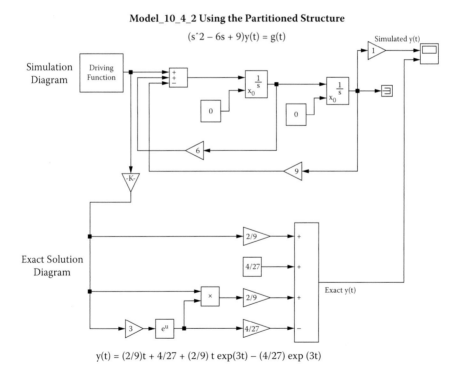

$$y(t) = (2/9)t + 4/27 + (2/9)\, t \exp(3t) - (4/27) \exp(3t)$$

FIGURE 10.9 The final Model_10_4_2 of the partitioned structure.

10.1.5 The Transfer Fcn Block from the Continuous Library

Returning to the concept of a system described by a transfer function, we now examine how these descriptions are used in practice. The transfer function of a linear system is defined as the ratio of the Laplace transform of the output to the Laplace transform of the input, with all initial conditions assumed to be zero (Dorf and Bishop 2008). Engineers will often describe a system just by giving its transfer function, and the system's behavior can be found by simulating its transfer function directly.

To aid in the simulation of a system from its transfer function, Simulink provides a block that simulates any transfer function that can be written as a ratio of polynomials. Known as the Transfer Fcn block, it is located in the Continuous library. The parameters of this block are the coefficients of the powers of s in the numerator and denominator of the transfer function (see Figure 10.10).

To see how we simulate a system when we are given its transfer function, let's analyze an example discussed by Matko, Zupancic, and Karba (1992) with the transfer function

$$H(s) = \frac{1}{s^2 - 6s + 9} \tag{10.46}$$

with an input $g(t) = 2t$. The coefficients of the powers of s are supplied to the Parameters window as vectors with the coefficient of the highest power present appearing on the left and the lowest on the right. Since the numerator is the polynomial $1 \times s^0$, the coefficient vector of the numerator is [1]. The polynomial of the denominator is $1 \times s^2 + (-6) \times s^1 + 9 \times s^0$, so the coefficient vector of the denominator is [1 – 6 + 9]. The simulation model is very easy to construct and is shown in Figure 10.11. This simulation produces the same results that we had from Model_10_1_1, as seen in Figure 10.12.

10.1.6 Examples of Transfer Function Simulation

To see the ease of simulation with the Transfer Fcn block, let's look at three interesting examples.

10.1.6.1 Impulse-Driven Spring

Let's consider a mass, M, suspended by a spring with spring constant, k, that is hanging in a uniform gravitational field. The mass will have as its

FIGURE 10.10   The Transfer Fcn block parameters.

Transfer Function Block Simulation of Example Transfer Function

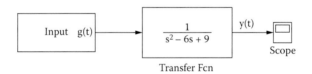

FIGURE 10.11   Model_10_5_1 for the transfer function of Equation (10.46).

normal, unperturbed configuration (called the normal operating position) the position $y(0) = 0$ and velocity $\dot{y}(0) = 0$. If we exert an impulse force on the mass, $r(t)$, that is a square impulse in the downward direction of amplitude -0.05 kg×m/s^2 and duration 0.1 s, then the mass will respond according to the dynamical Equation (10.47), assuming linear damping with a damping constant b.

$$M\frac{d^2 y(t)}{dt^2} + b\frac{dy(t)}{dt} + ky(t) = r(t)$$

(10.47)

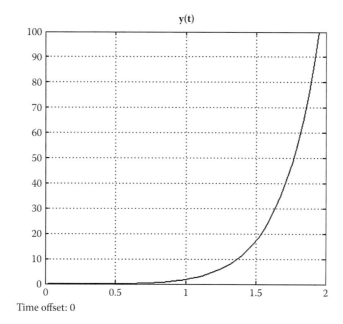

FIGURE 10.12   Results of Model_10_5_1. Values on the x-axis and y-axis are in arbitrary units.

If we find the Laplace transform of both sides with zero initial conditions, we find

$$Ms^2 y^*(s) + bsy^*(s) + ky^*(s) = r^*(s)$$ (10.48)

and the transfer function of this system is then given in Equation (10.49).

$$H(s) = \frac{\text{output}}{\text{input}} = \frac{y^*(s)}{r^*(s)} = \frac{1}{Ms^2 + bs + k}$$ (10.49)

Let's assume that $M = 1$ kg, $k = 25$ kg/s², and $b = 1$ kg/s. Then we have

$$H(s) = \frac{1}{s^2 + s + 25}$$ (10.50)

The model of this system is shown in Figure 10.13 and the results in Figure 10.14.

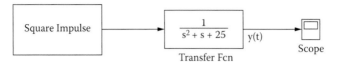

Mass on a Spring Subjected to a Square Impulse Force

FIGURE 10.13   Model_10_6_1 for the impulse response of a spring.

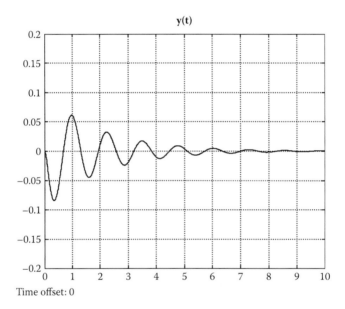

y(t)

Time offset: 0

FIGURE 10.14   The motion of a mass on a spring subjected to a square impulse. Displacement is shown on the *y*-axis in meters, and simulation time is shown on the *x*-axis in seconds.

10.1.6.2 Analog Signal Filter

An interesting area of application for simulation is the area of analog signal processing. Engineers in this area need to design a number of different signal-processing devices, so simulation becomes an excellent tool to use in testing their designs. One particular device is a filter that is used to remove unwanted frequencies from a signal.

A *lowpass filter* filters the frequencies higher than its cutoff frequency. Consider the transfer function for a second-order, lowpass Butterworth filter with a cutoff frequency of 50 Hz (Dorf and Bishop 2008).

Model_10_7_ for a Second Order, Lowpass Butterworth Filter with a 50 Hz Cutoff

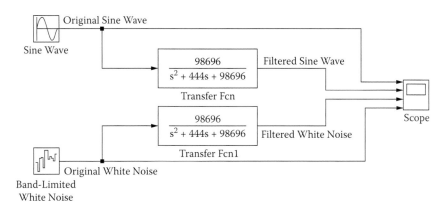

FIGURE 10.15 Model_10_7_1 of a second-order, lowpass Butterworth filter with 50-Hz cutoff. Frequency is shown on the *y*-axis in Hertz, and simulation time is shown on the *x*-axis in seconds.

$$H(s) = \frac{98696}{s^2 + 444s + 98696} \tag{10.51}$$

This transfer function has the simulation diagram in Figure 10.15, and the results for a pure sine wave of frequency 50 Hz combined with white noise are shown in Figure 10.16. As we can see, this filter is very effective in reducing the level of white noise by suppressing the higher frequencies in the white-noise component.

10.1.6.3 Disk Drive Motion under Control

Dorf and Bishop (2008) describe a model of a disk drive in a computer as a magnetic surface, a magnetic read/write arm, an arm motor, and an arm position sensor. When the arm is moved to another location on the surface, the motor receives a movement signal from the sensor until the sensor finds that the arm is positioned correctly. They supply a feedback diagram for the movement system similar to the one in Figure 10.17.

This diagram gives us a simulation model immediately if we have the transfer functions and input signals for the system. From Dorf and Bishop (2008), we can write the transfer functions for these elements, as seen in Table 10.2.

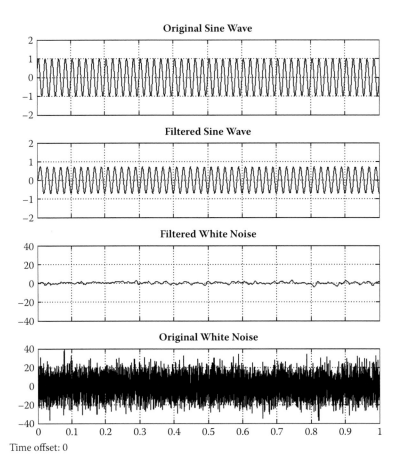

FIGURE 10.16 Second-order, lowpass Butterworth filter reductions.

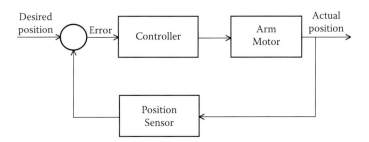

FIGURE 10.17 Feedback diagram of simple disk drive arm movement.

TABLE 10.2 Transfer Functions for the
Movement System Elements

Controller	K_a
Arm motor	$G(s) = \dfrac{K_m}{s(Js+b)(Ls+R)}$
Position sensor	$H(s) = 1$
Arm motor	$\dfrac{5}{s^2 + 20s}$
Arm sensor	$1/1$

Simple Disk Drive Arm Movement Model

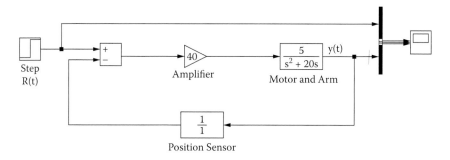

FIGURE 10.18   Model_10_8_1 for the disk drive arm movement.

If we use a value of 40 for the amplifier gain, we have Model_10_7_1 shown in Figure 10.18.

The results of our simulation in Figure 10.19 show how the disk head moves under a square input to a new address on the drive.

10.2 SIMULATION OF CHAOTIC DYNAMICS

When we are dealing with dynamical systems, we expect that they will be well behaved in the sense that a small change in initial conditions will produce a small change in the dynamics of the system, provided that we are not close to a singularity in the system. And most of the systems we have examined so far have generally agreed with this expectation. In this section, we consider systems that do *not* meet this expectation. We will find that they are so extremely sensitive to small changes in the initial conditions that they are essentially unpredictable in behavior and are called *chaotic* systems.

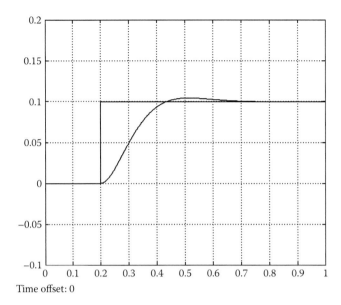

Time offset: 0

FIGURE 10.19 Simulation results for arm movement in the simple hard disk drive. Displacement is shown on the y-axis and simulation time on the x-axis in arbitrary units.

10.2.1 Example of a Chaotic System—the Lorenz Model

The premier example of a chaotic system is our weather system. In fact, attempts to produce accurate, predictable models of the weather led to the first descriptions of chaotic systems and the realization of how widespread they are. This description was first made by a meteorologist named Lorenz, who was able to produce a model of buoyant convection in a fluid with a 12-variable system of nonlinear ordinary differential equations. When he studied the system, he found aperiodic behavior that was extremely sensitive to initial conditions.

He was then able to isolate the phenomena to a simpler system of three variables: $x(t)$, the rate of convection overturn in the atmosphere; $y(t)$, the horizontal temperature gradient; and $z(t)$, the vertical temperature gradient. The equations describing this system were

$$\dot{x}(t) = \sigma\big(y(t) - x(t)\big) \tag{10.52}$$

$$\dot{y}(t) = rx(t) - y(t) - x(t)z(t) \tag{10.53}$$

The Lorenz Model for Buoyant Convection in a Fluid

sigma = 10, b = 8/3, r = 20, x(0) = 1, y(0) = 1, z(0) = 20

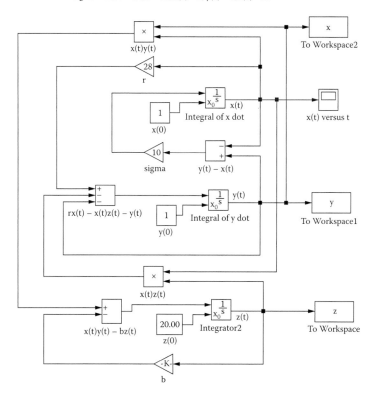

FIGURE 10.20   Simulation of the Lorenz model.

$$\dot{z}(t) = x(t)y(t) - bz(t) \tag{10.54}$$

where σ, r, and b are positive constants. This is called the Lorenz model and is the classic example of the chaotic system. The simulation in Figure 10.20 can be used to see the results he found.

The results for this system when we start with initial conditions of $x(0)$ = 1, $y(0)$ = 1, $z(0)$ = 20.00 and constants of σ = 10, r = 28, and b=8/3 are shown in Figure 10.21.

Now, if we change the initial condition for $z(0)$ to the value 20.01, we get the result in Figure 10.22.

Note that the evolution of $x(t)$ starting at about t = 6.0 is completely different for these two simulations, but the initial conditions have changed

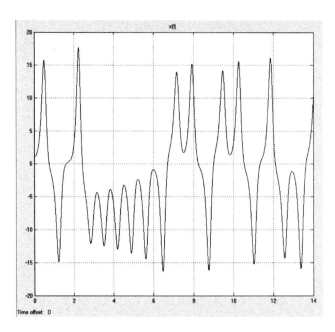

FIGURE 10.21   Simulation result for the Lorenz model when $z(0) = 20.00$.

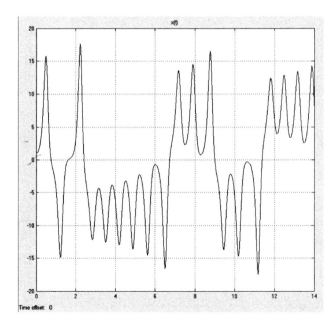

FIGURE 10.22   Simulation result for the Lorenz model when $z(0) = 20.01$.

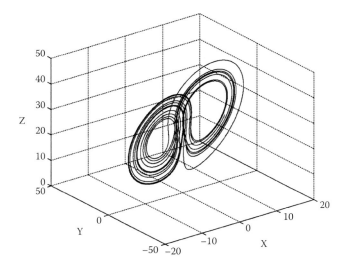

FIGURE 10.23 (See color insert following page 144.) Lorenz model simulation results in 3-D Cartesian space.

very little. This demonstrates the unpredictable nature of chaotic systems and the difficulty we have in using such models.

These results reveal a different picture, however, when we plot the x, y, z values at a given time as points in three-dimensional (3-D) Cartesian space. Figure 10.23 shows the plot.

Now the results look much more organized. They look like a complicated, but stable, aperiodic orbit about two points. Since the astronomical analog is very clear, the points are said to lie on an *attractor* trajectory. The center line is the attractor line, and it serves to separate the two trajectories by sending them to different orbits. The centers of the two orbits are two of the three steady states of the Lorenz model.

Chaotic behavior of a model arises in the nonlinear terms that appear in the dynamical equations. When we find nonlinear terms in a model, we must look for chaotic behavior in its dynamics.

10.2.2 The Logistic Map

A simpler example of a chaotic system is the *logistic map*. This surprisingly simple model of population development, which dates back to 1845, exhibits chaotic behavior that can be modeled quite easily.

$$x_{n+1} = 4r x_n (1 - x_n) \qquad x_i \in [0,1] \tag{10.55}$$

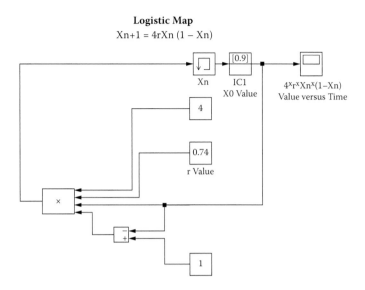

FIGURE 10.24   Simulation model for the logistic map.

Note that the dynamical equation is nonlinear (the right-hand side contains terms in x_n^2), so chaotic behavior is not unexpected. The source of the nonlinearity is a growth factor that decreases as the population increases, so that population limitation is captured in the model. The simulation of this model is given in Figure 10.24.

If we set the parameter $r = 0.5$ and $x_0 = 0.2$, we find that x stabilizes at the value 0.5 after an initial transient phase. The values for x_n are shown in Figure 10.25. Choosing different initial values for x_0 shows that $x_0 = 0.5$ is always the stable solution.

If we change the value of $r = 0.8$, however, we find that the stable solution changes into an oscillation between two values, 0.8 and 0.5, as we see in Figure 10.26.

As an aside, there is an interesting graphical method for understanding how these differences arise. The stable solution for x is achieved when the left-hand and right-hand sides of the logistic map are equal. Suppose we plot the left-hand and right-hand sides as functions on the same graph. For the case when $r = 0.5$, we find the graph shown in Figure 10.27.

If we start by finding what the value of the right-hand side will be when x has its initial value, we must draw a vertical line from $x_0 = 0.2$ to the parabola, producing point A. Then, to set the left-hand side equal to that value, we must draw a horizontal line to the straight line, producing point B. Now we must repeat the process, but this time we start the right-

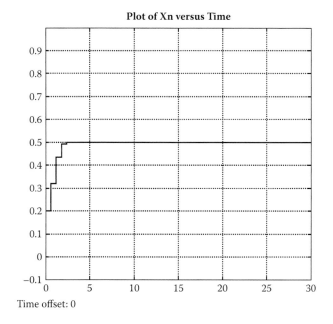

FIGURE 10.25 Stable solution for the logistic map.

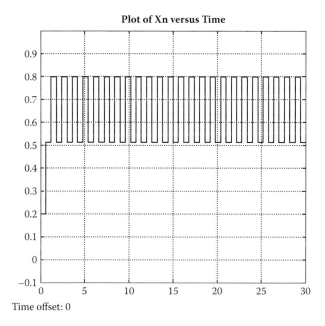

FIGURE 10.26 Two-valued stable solutions for the logistic map.

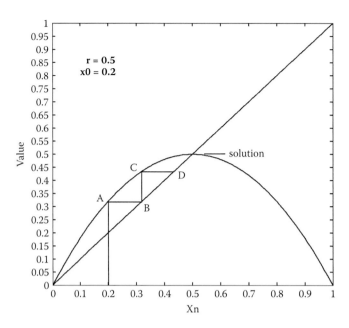

FIGURE 10.27   Left-hand and right-hand sides of the logistic map with $r = 0.5$ and $x_0 = 0.2$.

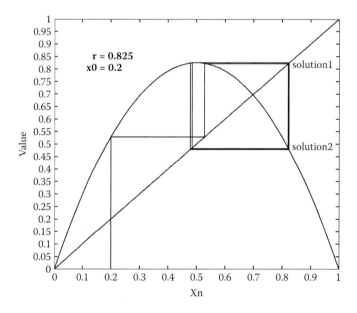

FIGURE 10.28   Left-hand and right-hand sides of the logistic map with $r = 0.825$ and $x_0 = 0.2$.

hand-side computation with the value from point B, producing point C. Extending to the horizontal line, we produce point D. If we continue this process, the two lines will converge to point X, which will be the value at which both sides are equal. This is the stable solution, and we can see that it produces the same value 0.5 that we found in our simulation.

Now what happens when we change the value of r to 0.825? The right-hand side of the equation will rise, causing the graph in Figure 10.28.

Repeating the same process as before, we find that we arrive at a box that repeats itself forever, producing two values that alternate. This is exactly what we found from the simulation. It shows that the single-valued stable solution bifurcates into a two-valued stable solution as r increases, giving the values 0.82 and 0.48.

If we continue to increase the value of r slowly, we find that the number of stable solutions continually grows, as we see in Figure 10.29, until we find so many solutions that the behavior can only be regarded as chaotic. Trying to predict the dynamics of the system in this range of r values is simply not possible.

The usual way of displaying this behavior is to plot the values of the stable solutions on the y-axis and the values of r on the x-axis of a graph

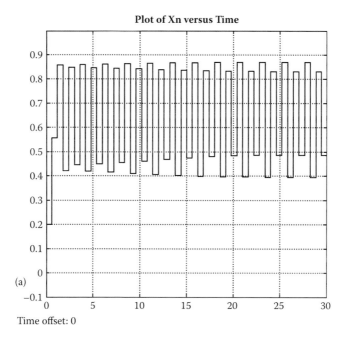

Time offset: 0

FIGURE 10.29   Stable solutions for $r = 0.87$.   (Continued)

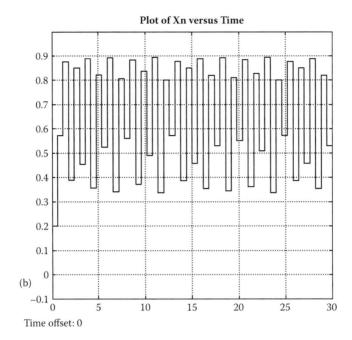

(b)

Time offset: 0

FIGURE 10.29 (Continued)   Stable solutions for $r = 0.895$.

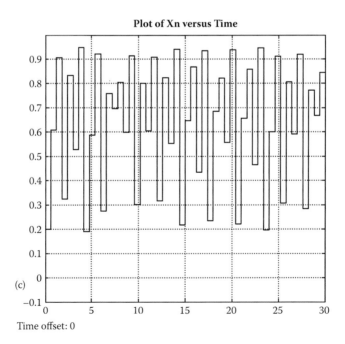

(c)

Time offset: 0

FIGURE 10.29   Stable solutions for $r = 0.95$.   (*Continued*)

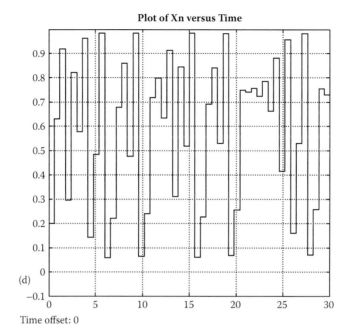

FIGURE 10.29 (Continued)   Stable solutions for $r = 0.9855$.

called a *bifurcation graph*. Such a graph can be produced easily in the MATLAB® environment, but not as well inside Simulink®. The interested reader should consult Higham and Higham (2005) for explicit M-file language for producing such graphs.

EXERCISE 10.1

Using the logistic map, described by the equation

$$x_{n+1} = 4 r x_n (1 - x_n) \qquad (10.56)$$

1. Construct a Simulink model that can be used to demonstrate that this system has a stable, single-valued, fixed-point solution when $x_0 = 0.1$ and $r = 0.7$. What is the value of this fixed-point solution? Give the answer to an accuracy of two decimal places.
2. If the parameter r is increased, the solution will eventually begin to bifurcate. At what value of r will the system begin to exhibit a stable, four-valued, fixed-point solution? Give the answer to an accuracy of two decimal places.

10.3 SIMULATION OF PARTIAL DIFFERENTIAL EQUATIONS

In this section, we introduce the simulation of systems whose behavior is described by continuous functions of *multiple* independent variables. These systems comprise a large part of the issues that scientists and engineers deal with. The dynamics of these systems is more difficult than we have seen in our study of systems described by ordinary differential equations. They have models that are usually described in the form of dynamical equations called *partial differential equations* (PDEs). Since the study of partial differential equations is an advanced topic and can only be introduced very briefly at this level, our real focus in this section is to introduce the Embedded MATLAB Function block. We use this Simulink feature to produce a block that can solve a system's dynamics when it is described by a specific partial differential equation.

10.3.1 What Is a Partial Differential Equation?

Suppose we have a function of multiple independent variables and we want to know how this function changes when only one of the variables is changed while the others are held constant. The answer is given by a partial derivative of the function with respect to the independent variable allowed to change. Partial derivatives are notated in a way similar to ordinary derivatives. The ordinary derivative of a single-variable function $f(t)$ is written as

$$\frac{df(t)}{dt}$$

while the partial derivative of a multivariable function $f(x,y,z,t)$ with respect to the t variable is written

$$\frac{\partial f(x,y,z,t)}{\partial t}$$

and with respect to the x variable is written

$$\frac{\partial f(x,y,z,t)}{\partial x}$$

and so on.

As we have seen throughout this book, the fundamental equations describing the dynamical behaviors of systems are formed by relating ordinary derivatives of the system's fundamental functions and parameters to functions of the single independent variable, t. But in this section, we turn to systems whose dynamical equations must relate partial derivatives of the system's fundamental functions and parameters to functions of the multiple independent variables. These equations are called partial differential equations.

> A partial differential equation is an equation containing partial derivatives of quantities of interest in the system with respect to one or more independent variables of the system.

The order, degree, and linearity notation used for ordinary differential equations are used in the same way for partial differential equations. As a result, partial differential equations look very familiar to us. Unfortunately, the mathematics for solving partial differential equations analytically is not as easy. While there is a general, systematic procedure by which every ordinary differential equation obeying certain conditions can be solved exactly, there is no such general procedure for partial differential equations. Instead, PDEs are grouped into classes of equations based on their forms, and these categories are attacked by different analytic methods.

Just as we saw with ordinary differential equation systems, a set of initial conditions is required to determine a unique solution. In the ordinary differential equation systems, these were the value of the function and some of its derivatives at $t = 0$, but here we have more possibilities for initial conditions. We could have the value of the function and some of its derivatives at $t = 0$, which we would call *initial values*. But we could also have the value of the function and its derivatives at particular geometric points that are constant in time. These initial conditions would be called *boundary values*. Of course, we could also have a mixture of both kinds of initial conditions.

10.3.2 Examples of Systems with Partial Differential Equation Models

A classic example of a partial differential equation model is the heat-diffusion problem. Suppose we have a thin rod with an initial temperature distribution along its length and we want to know how the temperature will change in time. Since the temperature at a point in the rod depends on both the location of the point and the time, this is a two-variable problem. The dynamical equation describing the evolution of temperature is the two-variable heat-diffusion equation (Garcia 2000), which is a partial differential equation.

$$\lambda \frac{\partial^2}{\partial x^2} T(x,t) - \frac{\partial}{\partial t} T(x,t) = 0 \tag{10.57}$$

In this equation, κ, called the heat-diffusion coefficient, is a constant describing the rod material's thermal properties. There are other systems with similar partial differential equations for their dynamics, and these kinds of partial differential equations are called *parabolic* equations.

The initial conditions for the heat-diffusion problem are normally a mixed set of initial and boundary values. We would expect to have the initial values of the temperature distribution along the rod and also have the boundary values of the temperatures or the heat flow at the ends of the rod.

A second example of partial differential equation models includes systems with dynamics determined by the *wave equations*. The propagation of a pressure wave through a one-dimensional fluid is described by the two-variable equation

$$v^2 \frac{\partial^2}{\partial x^2} p(x,t) - \frac{\partial^2}{\partial t^2} p(x,t) = 0 \tag{10.58}$$

where the parameter v is the speed of the pressure propagation in the fluid. This equation has a different form from the heat-diffusion equation. The partial differential equations that have this form are called the *hyperbolic* equations. The initial conditions for this problem are similar to the heat-diffusion problem. We could be given the shape of the wave at $t = 0$ and then be asked to find the shape for all $t \geq 0$. The wave-equation problem is more likely to be posed over an interval of fixed length, L, and have boundary values called *periodic* boundary values, in which

$$p(0,t) = p(L,t), \quad \frac{\partial}{\partial x} p(0,t) = \frac{\partial}{\partial x} p(L,t)$$

and so on.

A third example of partial differential equation models includes systems determined by Poisson's equation, which have two independent variables, neither of which is time. Although this is not a dynamical system, it is a classic example of a partial differential equation model. This equation occurs when it is necessary to determine the electric field at a point in a plane due to a fixed distribution of electric charge in the plane.

$$\frac{\partial^2}{\partial x^2}V(x,y)+\frac{\partial^2}{\partial y^2}V(x,y)=-\frac{\rho(x,y)}{\varepsilon_0} \tag{10.59}$$

In this equation, $V(x,y)$ is the electric potential, $\rho(x,y)$ is the charge density distribution, and ε_0 is a constant for the electromagnetic field. These kinds of partial differential equations are called *elliptical* equations.

10.3.3 Simulating Partial Differential Equation Models

How can we simulate the dynamics of systems that are described by partial differential equations? By now, our first reaction might be to go to the Simulink libraries and look for a Partial Integrator block to use, just as we did for the ordinary differential equation models. But the only block we have is an ordinary differential equation Integrator block that can be used only for functions of a single variable. So we must take a different approach; we must produce an algorithm for solving the partial differential equation and then create a block of our own to implement the algorithm. Let's look first at how we can produce such an algorithm.

We'll use the heat-diffusion problem in a thin rod as the infrastructure on which we'll demonstrate this approach. We'll ask for the temperatures at 20 evenly spaced points along the rod, with the left end being point 1 and the right end being point 20. The thin rod has initial conditions: (a) the ends of the rod are fixed at a temperature of 0°C for all times and (b) at $t = 0$, the temperatures at the points 2,3,...,11 along the rod increase by increments of +1°C up to point 11, which has a temperature of 9°C that decreases thereafter at the points 12,13,...,19 by decrements of −1°C. We recognize that these initial conditions are a mixed set of initial and boundary values.

The equation we must find an algorithm for is Equation (10.59). For purposes that will become clear later, this equation is rewritten as

$$\frac{\partial}{\partial t} T(x,t) = \sigma \frac{\partial^2}{\partial x^2} T(x,t) \qquad (10.60)$$

Let's transform the partial derivatives on each side of this equation into finite difference expressions so that we can use the difference-equation approach we studied in Chapter 3.

To do this we break both time and space into finite intervals. We use a basic time step, τ, just as in the finite difference models, replacing t with $n\tau$ for $n = 0,1,2,\ldots$. Now we assume that the thin bar has length L, and we break it up into μ intervals of size

$$l = \frac{L}{m}$$

replacing x with ml for $m = 0,1,2,\ldots,\mu$. We use the same notation that we used in the difference-equation systems. Using a Taylor's expansion of the function $T(x,t)$ in the same way we did in Chapter 4, we can find two expansions that enable us to find an expression for the partial derivatives: a first derivative of the function at time $n + 1$ holding x constant and a second derivative at the points $m - 1$ and $m + 1$ holding t constant. This gives us

$$\frac{\partial}{\partial t} T(x,t) \rightarrow \frac{T_{n+1}^m - T_n^m}{\tau} \qquad (10.61)$$

$$\frac{\partial^2}{\partial x^2} T(x,t) \rightarrow \frac{T_n^{m+1} + T_n^{m-1} - 2T_n^m}{l^2} \qquad (10.62)$$

Substituting Equations (10.61) and (10.62) into Equation (10.63) gives us the algorithm we need to compute the next time value for each spatial point on the bar.

$$T_{n+1}^m = T_n^m + \frac{\kappa\tau}{l^2}\left(T_n^{m+1} + T_n^{m-1} - 2T_n^m\right) \qquad (10.63)$$

What this algorithm tells us is that the value of the temperature at a point on the rod is determined by the previous value plus an amount that is determined by the previous values at the points on either side, the thermal properties of the rod, and the step size used. This is quite reasonable, since the partial differential equation states that the rate of change of the temperature at a point is proportional to the rate of spatial change along the rod. We expect that the values of the temperature on the sides of the point will be necessary to perform this calculation. We also note that the values of the temperatures on the right-hand side are all functions of the nth time step, so this is an explicit method of order 1 using the terminology of Chapter 5.

We now must build a Simulink diagram that uses the algorithm above to solve the heat diffusion equation. To do this, we need to solve two problems: (a) a mechanism to save and reuse all the temperatures along the rod, and (b) a mechanism to use the saved temperatures to compute the next set of temperatures along the rod. To solve the first problem, we'll use Simulink vector variables to hold all the temperature values for a time step. To solve the second problem, we'll use the embedded MATLAB Function block from the User-Defined Functions library to define our own user-defined Simulink block to perform the calculations needed by Equation (10.63).

10.3.4 The Embedded MATLAB Function Block from the User-Defined Functions Library

This block enables a simulator to create a block containing a function definition that will be evaluated when the block is executed by Simulink. The evaluation of the function is actually done in the underlying MATLAB environment, and the result of the evaluation is returned by MATLAB to Simulink so that it becomes available for use in the Simulink environment. The block has a single input port on which it receives a standard Simulink input, which could be a vector or matrix input. It has a single output port on which it outputs its result. So it acts just like any other Simulink block in the way it is used in constructing a simulation. It has few parameters, as we see in Figure 10.30.

The key to using this block, of course, is what we put inside it. If the block is double-clicked, an editor opens up containing a template function definition that enables the simulator to define the action of the block by using a MATLAB function. This editor is shown in Figure 10.31.

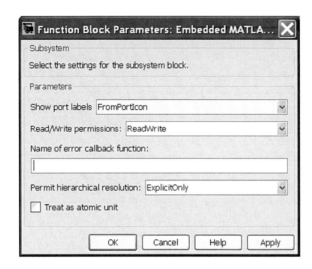

FIGURE 10.30   The embedded MATLAB function block parameters.

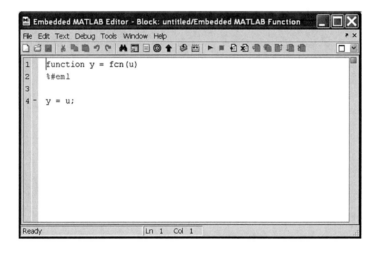

FIGURE 10.31   The embedded MATLAB function block editor.

Note that the editor starts the function creator with a skeleton function template. In the template, we see that the input and output ports have standard variable names: u for input and y for output. Also notice that the input port supplies the input to the function through the variable u, and MATLAB outputs the result of function evaluation through the variable y to the output port. The single line y = u defines the default function to be evaluated. All this line does is pass the input through to the output.

We don't have to do much in the writing of code to get this apparatus to function for us. All we have to do is to replace the default line by one or more lines that define the function we would like to have evaluated, making sure that the input is named u and the result is named y. We may also want to add additional inputs, and Simulink will automatically add new input ports for these inputs. The challenge is to write the function-defining lines in proper MATLAB code so that MATLAB will be able to understand the desired function.

Since we are just conducting an overview of partial differential equation models and their simulation in this section, we won't study the MATLAB language that is used for function construction. Instead, the interested reader should refer to Moler's introduction (2004) for the complete discussion. Here, we simply show the MATLAB function definition that will compute Equation (10.63) for us.

To define the function, we add one new input parameter to the block to supply the coefficient κ needed in Equation (10.63). This requires that we change the first line of the function to

$$\texttt{function y = fcn(u,coeff)} \tag{10.64}$$

We need to insert a statement to find the number of spatial steps along the rod. The following line instructs MATLAB to determine the number of temperatures in the vector by using its built-in `length(u)` function to determine this number. Computing the number in this way makes our function definition reusable for any value of m.

$$\texttt{m = length(u)} \tag{10.65}$$

Next, we must initialize the output vector by copying the input vector components to it. The variable u is the input that contains the vector of temperatures at the m points along the thin rod. So the following line copies the input vector into the output vector so that we have the correct number of output values in y.

$$\texttt{y = u} \tag{10.66}$$

Now we insert the assignment statement that causes each output component to be changed to the value required by Equation (10.63).

```
y(2:(m-1))=u(2:(m-1))+coeff*(u(3:m)+u(1:(m-2))-2*u(2:(m-1))
```
$$(10.67)$$

In this line, we see why we need to compute m. We need to use it to instruct MATLAB which of the values of y we need to recompute so that y has the new values of temperature along the rod. The `y(2:(m-1))` notation is MATLAB's way of indicating that y contains a vector of values from 2 to m-1 and that we want to recompute it, component by component, starting with component 2 and ending with component m-1 (which will be 19 for the 20-step rod). So, MATLAB will start off by computing the new value for the first of this sequence of components `y(2)` using the equation

```
y(2)=u(2)+coeff*(u(3)+u(1)-2*u(2))
```
$$(10.68)$$

This equation takes the input value of component 2 and adds to it the result of "averaging" the two input values of the components on each side of it (component 1 and component 3), with itself using the specified weightings. Remember that the input value for a component is the value computed for the component in the *previous* time step (time step *n* in Equation [10.63]). This is exactly what Equation (10.63) tells us to do to find the new value for the m=2 point on the rod. Next, MATLAB will compute `y(3)` using

```
y(3)=u(3)+coeff*(u(4)+u(2)-2*u(3))
```
$$(10.69)$$

This iteration will end at y(19) with

```
y(19)=u(19)+coeff*(u(20)+u(18)-2*u(19))
```
$$(10.70)$$

We may wonder here why we didn't recompute the temperatures at `y(1)` and `y(20)`. Leaving these two points with a value of 0 is our way of satisfying the boundary values that say that the ends of the rod are 0 for all times. If we don't change them, the condition is met.

Entering these lines into the Function Editor window gives us the full function definition, as shown in Figure 10.32. Note that we've added documentation so that the function can be understood by simulators who need to maintain or modify it.

Let's enter a triangular initial temperature distribution for the rod to see the simulation execution. We take the sequence of temperatures starting

```
Embedded MATLAB Editor - Block: Model_12_1_1/Embedded MATLAB Function1
File  Edit  Text  Debug  Tools  Window  Help
1    function y = fcn(u, coeff)
2    % This block supports an embeddable subset of the MATLAB language.
3    % See the help menu for details.
4
5    % Michael A. Gray, August 17, 2009.
6
7    % This function uses the algorithm
8    %
9    % T(m,n+1) = T(m,n) + (Kappa*Tau/h^2)(T(m+1,n) + T(m-1,n) - 2*T(m,n))
10   %
11   % from Numerical Methods for Physics, Second Edition, by Alejandro
12   % L. Garcia, Chapter 6, to solve the heat diffusion equation for a
13   % thin rod of length 1 having 20 equally-spaced points along the rod.
14   %
15   % The initial conditions are
16   %
17   % (1) the ends of the rod are fixed at a temperature of 0 for all times
18   % (2) at  t = 0, the temperatures at the points along the rod are input
19   %     in the initial input vector u.
20
21   m = length(u);
22   y = u;
23   y(2:(m-1)) = u(2:(m-1))+coeff*(u(3:m)+u(1:(m-2))-2*u(2:(m-1)));

Ready                    Ln 1    Col 1
```

FIGURE 10.32   The final function definition in the embedded MATLAB function block.

with point 0 to be [0 0 1 2 3 4 5 6 7 8 9 8 7 6 5 4 3 2 1 0]. This is a triangular distribution with a maximum of 9°C at point 10 and minima of 0°C at points 0 and 20. The difference-equation model for the thin-rod simulation is shown in Figure 10.33. Executing the model with a fixed step size of 0.0001 and ode4 (the *Runge–Kutta* fixed-step solver) produces the Scope output in Figure 10.34. Note that the calculation subsystem for the coefficient

$$\frac{\kappa\tau}{l^2}$$

requires that the fixed step size used in the configuration parameters must also be entered into the constant τ inside the subsystem.

In the Scope output, we have one curve for each temperature point. In the actual window, these are distinguished by the color assignment we discussed in Chapter 3.

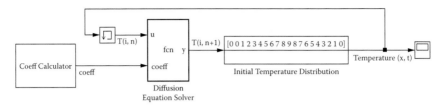

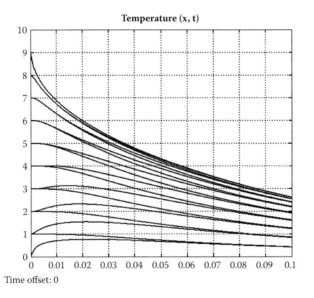

FIGURE 10.33 Model_10_1_1 for heat diffusion in a thin rod.

FIGURE 10.34 (See color insert following page 144.) The temperature dynamics for 20 points in a thin rod.

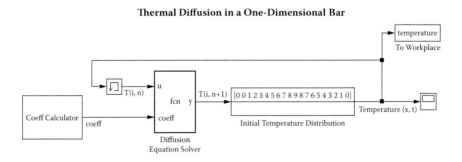

FIGURE 10.35 The heat diffusion in a thin rod modified to write to the MATLAB environment.

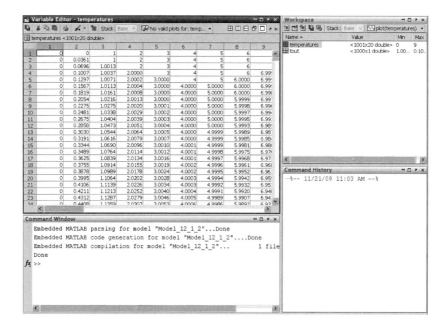

FIGURE 10.36 The temperatures in the MATLAB environment after simulation.

While the Scope data is correct, it does not reveal the qualitative features of the simulation in a striking way. So let's make a new model that writes the results to the MATLAB environment using the To Workspace block. Model_10_1_2 contains this modification, as shown in Figure 10.35.

If we rerun the model, open the MATLAB window, and open the array variable `temperatures` in the MATLAB Array Editor, we then see the array of temperature values, as shown in Figure 10.36.

The `temperatures` variable is a 1001 × 20 array, with the rows containing the temperature distribution at different time steps and the columns containing the temperatures for points on the rod. These values are stored in temperatures exactly as needed to produce a three-dimensional plot in MATLAB. If we execute the command `mesh(temperatures)`, we see the revealing plot in Figure 10.37. This plot shows how temperature versus distance along the rod develops in time. Initially, the temperatures form linear lines from the outside ends of the rod to a peak in the center. But as time elapses, the linear lines diffuse into a smooth curve, reaching a flat line of 0's at a future time.

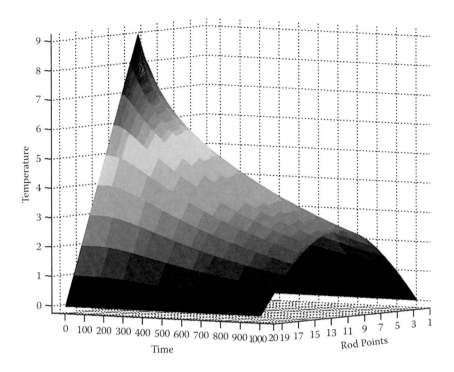

FIGURE 10.37 (See color insert following page 144.) The temperature–time surface for the thin rod.

EXERCISE 10.2

Simulate the temperature diffusion in a thin rod, finding the temperatures, $T(x_m, t_n)$, at 20 evenly spaced points along the rod, with the left end being point 1 and the right end being point 20. Use the initial conditions given in Table 10.3.

Plot the temperature–time surface using the MATLAB mesh function.

10.4 SUMMARY

Transforming dynamical equations into alternative spaces can sometimes make the simulation problem much easier. The Laplace transform is the primary transform used for this purpose, and it transforms a differential equation in t-space into a polynomial equation in s-space. After solution in s-space, the solution is inverted to give the solution in t-space.

Simulations using the Laplace transform can be organized into different structures called nested or partitioned structures. The Integrator block simulates the $1/s$ terms in a transform directly, so the use of Integrator

TABLE 10.3   Initial
Temperatures for the Points in
Exercise 10.2

$T(1,t_n)$	0 for all $t_n \geq 0$
$T(2,0)$	0
$T(3,0)$	1
$T(4,0)$	2.2
$T(5,0)$	3.6
$T(6,0)$	5.4
$T(7,0)$	7.4
$T(8,0)$	9.4
$T(9,0)$	10.4
$T(10,0)$	11.4
$T(11,0)$	12
$T(12,0)$	11.4
$T(13,0)$	10.4
$T(14,0)$	9.4
$T(15,0)$	7.4
$T(16,0)$	5.4
$T(17,0)$	3.6
$T(18,0)$	2.2
$T(19,0)$	1
$T(20,t_n)$	0 for all $t_n \geq 0$

blocks makes the construction of these structures relatively simple. Even easier to use is the Transfer Fcn block that allows for easy simulation of transfer functions expressed as ratios of polynomials in s-space. When a transfer function is known for a complex system, this block makes it easy to construct a simulation directly from the transfer function.

In this chapter, we also studied a class of systems whose dynamics are very difficult to simulate because of the nonlinear nature of their models. These nonlinearities lead to extreme sensitivity to small changes in parameter values, which causes the solutions to change wildly. Such systems are called *chaotic systems*, and their dynamics must be analyzed through phase space methods to get a correct picture of their dynamics.

The dynamical behavior of many systems is described by continuous functions of *multiple* independent variables. The equations used for the important functions or parameters of these continuous systems are called partial differential equations (PDEs). A partial differential equation is an equation containing partial derivatives of quantities of interest in the system with respect to one or more independent variables. A partial derivative is notated by the symbol

$$\frac{\partial}{\partial v}$$

where v is the independent variable. The order, degree, and linearity notation used for ordinary differential equations are used in the same way for partial differential equations. The initial conditions for partial differential equations can be a combination of initial values and boundary values. There is no general solution procedure for partial differential equations, which increases the importance of simulation for analyzing systems of this kind.

When we examined Simulink to see how to perform a simulation of partial differential equations, we found that Simulink does not provide solvers for these equations. However, these systems can be simulated using MATLAB by producing an algorithm for the PDE and creating a user-defined MATLAB function to implement the algorithm. The embedded MATLAB function block can be used to create a user-defined solver block, which allows us to use the embedded MATLAB function block to write our own solvers for the models in the cases where we have algorithms for solving these equations. The resulting values can be retransmitted to the underlying MATLAB environment by using the To Workspace block. Then three-dimensional plots can be produced from the stored variables by using the MATLAB 3-D plotting functions, such as surf and mesh.

REFERENCES AND ADDITIONAL READING

Dorf, R. C., and R. Bishop. 2008. *Modern Control Systems.* Upper Saddle River, NJ: Prentice Hall.

Garcia, A. 2000. *Numerical Methods for Physics.* Upper Saddle River, NJ: Prentice Hall.

Gershenfeld, N. 1999. *The Nature of Mathematical Modeling.* Cambridge: Cambridge University Press.

Gould, H., and J. Tobochnik. 1996. *An Introduction to Computer Simulation Methods: Applications to Physical Systems.* New York: Addison Wesley.

Higham, D. J., and N. Higham. 2005. *MATLAB Guide.* Philadelphia: Society for Industrial and Applied Mathematics.

Klee, H. 2007. *Simulation of Dynamic Systems with MATLAB and Simulink.* Boca Raton, FL: CRC Press.

Matko, D., B. Zupancic, and R. Karba. 1992. *Simulation and Modeling of Continuous Systems: A Case Study Approach*. New York: Prentice Hall.

Moler, C. B. 2004. *Numerical Computing with MATLAB*. Philadelphia: Society for Industrial and Applied Mathematics.

Severance, F. 2001. *Systems Modeling and Simulation: An Introduction*. New York: John Wiley.

The MathWorks, Inc. 2007. http://www.mathworks.com.

Appendix A: Alphabetical List of Simulink Blocks

Block Name	Library Name	Book Section
Add	Math Operations	2.5.2.2
Bus Creator	Signal Routing	6.7.1
Clock	Sources	2.5.1
Constant	Sources	2.3.2.1
Display	Sinks	3.6.2
Embedded MATLAB Function	User-Defined Functions	10.3.4
Floating Scope	Sinks	3.4.1
Gain	Math Operations	2.7.2
IC	Signal Attributes	3.3.4
If	Ports & Signals	7.4.1.4
If Action Subsystem	Ports & Subsystems	7.4.1.3
Integrator	Continuous	4.4.1
Math Function	Math Operations	2.8.1
Memory	Discrete	3.3.2
Merge	Signal Routing	7.4.1.5
Product	Math Operations	2.5.2.1
Scope	Sinks	2.3.1.1
Sine Wave	Sources	2.7.1
Subsystem	Ports & Subsystems	3.5.1
Switch	Signal Routing	7.4.1.1
Terminator	Sinks	9.1.2
To Workspace	Sinks	4.5.1
Transfer Fcn	Continuous	10.1.5
XY Graph	Sinks	2.7.3
Zero-Order Hold	Discrete	7.4.1.2

Appendix B: The Basics of MATLAB for Simulink Users

MATLAB® is a large scientific, engineering, and applied mathematical computation system widely used in industry and academia. The system is much too large and complex for us to describe in this short appendix, so we just refer the reader of the following discussion for more details on the system and note particularly that Attaway's book (2009) is an excellent introduction for a MATLAB beginner. In this appendix, we only consider a few things that are helpful and relevant to the Simulink® user.

B.1 THE MATLAB MAIN WINDOW

To start Simulink, we must first start MATLAB. When MATLAB is first executed, it displays a main window that has horizontal toolbars at the top, with the rest of the window divided into three columns or subwindows, as shown in Figure B.1.

Two of these are arranged vertically on the window's right, and the other occupies the window's left. The upper pane on the right is typically referred to as the Workspace. The Current Directory tab will cause a list of the MATLAB-relevant files to appear, while the Workspace will display the current MATLAB variables residing in the Workspace.

The lower pane on the right shows the Command History, which is a list of all the commands executed by the user in the MATLAB Command Window since the execution began. Note that the Command History can span sessions, so it can accumulate a considerable record.

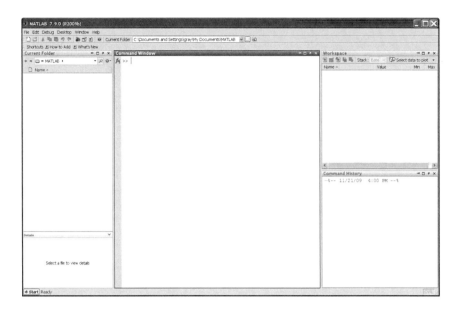

FIGURE B.1   The MATLAB main window.

The pane in the middle is the Command window, in which the user types MATLAB expressions for evaluation. Expression evaluation is how the user employs MATLAB to achieve the desired computations.

For example, if a user has defined two variables times and solution, which contain, respectively, a vector of time values and a vector of solution values, then the user can ask MATLAB to plot these by entering the command plot (times, solution) at a prompt in the Command window.

An important question for Simulink users is the location of the folder where files are placed by MATLAB, since Simulink stores its models in this folder unless the user has changed the default location.

B.2 THE DEFAULT FOLDER IN MATLAB

To display the location of all the folders used by MATLAB, the user can select the File|Set Path item in the File menu. The Set Path window is then displayed, as seen in Figure B.2 and the user can determine the location of the default folder.

The top folder is the default folder, so following the path to this folder brings the Simulink user to the folder where the models are stored by

FIGURE B.2   The MATLAB Set Path window.

default. This window also allows the Simulink user to change the location of the default folder. See the MATLAB documentation for the details.

B.3 LAUNCHING SIMULINK FROM MATLAB

After setting the location of the folder where the Simulink models are to be saved, the next step is to launch Simulink. The program can be launched most simply by selecting the Simulink icon on the toolbar. It consists of a red clock face with several shapes behind it. It looks somewhat like a cartoon farm tractor with large red wheels, as seen in Figure B.3 (look for the cursor to locate the icon).

Once Simulink is started, the user can minimize the MATLAB window for Simulink work. The MATLAB window cannot be closed, however, since that action closes the Simulink software as well.

B.4 SIMULINK MODEL FILES IN MATLAB

The last item discussed in this Appendix is the makeup of the Simulink model file. When a user is finished with a Simulink model, it can be saved with the File|Save or Save As menu item from the File menu on the

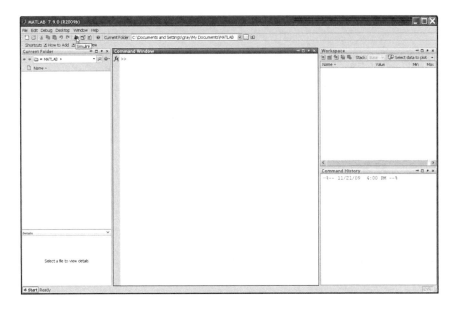

FIGURE B.3　(See color insert following page 144.) Location of the Simulink icon on the MATLAB toolbar.

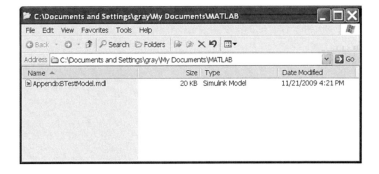

FIGURE B.4　The test Simulink model file in the default MATLAB folder.

Simulink Model window toolbar. This command causes Simulink to save the model file with identifier .mdl in the default folder. Figure B.4 shows the contents of the default folder after a Simulink model has been saved.

The last point for discussion in this Appendix is the format of the Simulink model file itself. This file is written to the folder as an ASCII text file, so it is easily viewed by any text editor. The format and content of the file must be maintained, however, so any text editor that inserts editor information should not be used to view the file. Figure B.5A shows a

```
AppendixBTestModel.mdl - WordPad
File  Edit  View  Insert  Format  Help

Model {
    Name                    "AppendixBTestModel"
    Version              7.4
    MdlSubVersion           0
    GraphicalInterface {
        NumRootInports          0
        NumRootOutports         0
        ParameterArgumentNames  ""
        ComputedModelVersion    "1.1"
        NumModelReferences      0
        NumTestPointedSignals   0
    }
    SavedCharacterEncoding  "windows-1252"
    SaveDefaultBlockParams  on
    ScopeRefreshTime        0.035000
    OverrideScopeRefreshTime on
    DisableAllScopes        off
    DataTypeOverride        "UseLocalSettings"
    MinMaxOverflowLogging   "UseLocalSettings"
    MinMaxOverflowArchiveMode "Overwrite"
    MaxMDLFileLineLength    120
    Created                 "Sat Nov 21 16:17:38 2009"
    Creator                 "gray"
    UpdateHistory           "UpdateHistoryNever"
    ModifiedByFormat        "%<Auto>"
    LastModifiedBy    "gray"
    ModifiedDateFormat      "%<Auto>"
    LastModifiedDate        "Sat Nov 21 16:21:18 2009"
    RTWModifiedTimeStamp    180721118
    ModelVersionFormat      "1.%<AutoIncrement:1>"
    ConfigurationManager    "None"
    SampleTimeColors        off
    SampleTimeAnnotations   off
    LibraryLinkDisplay      "none"
    WideLines          off
    ShowLineDimensions      off
    ShowPortDataTypes       off
    ShowLoopsOnError        on
    IgnoreBidirectionalLines off
    ShowStorageClass        off
    ShowTestPointIcons      on
    ShowSignalResolutionIcons on
    ShowViewerIcons    on
    SortedOrder             off
    ExecutionContextIcon    off
    ShowLinearizationAnnotations on
    BlockNameDataTip        off
    BlockParametersDataTip  off
    BlockDescriptionStringDataTip      off

For Help, press F1                                          NUM
```

FIGURE B.5A   Contents of the test Simulink model file from the default folder.

Notepad display of part of the contents of a test model file. The first window shows the start of the next window, and the right file shows the end of the file.

The end of the file shows that there are entries in the file for two blocks and one connection. One block is a Constant block, while the other is a Scope block. The connection starts at the Constant block and ends at the Scope block.

```
AppendixBTestModel.mdl - WordPad
File  Edit  View  Insert  Format  Help

System {
    Name                    "AppendixBTestModel"
    Location                [480, 104, 1016, 386]
    Open                    on
    ModelBrowserVisibility  off
    ModelBrowserWidth       200
    ScreenColor             "white"
    PaperOrientation        "landscape"
    PaperPositionMode       "auto"
    PaperType               "usletter"
    PaperUnits              "inches"
    TiledPaperMargins       [0.500000, 0.500000, 0.500000, 0.500000]
    TiledPageScale          1
    ShowPageBoundaries      off
    ZoomFactor              "100"
    ReportName              "simulink-default.rpt"
    SIDHighWatermark        2
    Block {
      BlockType             Constant
      Name                  "Constant"
      SID                   1
      Position              [125, 104, 155, 136]
      OutDataType           "fixdt(1, 16)"
      OutScaling            "2^0"
    }
    Block {
      BlockType             Scope
      Name                  "Scope"
      SID                   2
      Ports                 [1]
      Position              [300, 104, 330, 136]
      Floating              off
      Location              [188, 390, 512, 629]
      Open                  off
      NumInputPorts         "1"
      List {
        ListType            AxesTitles
        axes1               "%<SignalLabel>"
      }
      DataFormat            "StructureWithTime"
      SampleTime            "0"
    }
    Line {
      SrcBlock              "Constant"
      SrcPort               1
      DstBlock              "Scope"
      DstPort               1
    }
  }
}

For Help, press F1                                                      NUM
```

FIGURE B.5B   Contents of the test Simulink model file from the default folder.

REFERENCES AND ADDITIONAL READING

Attaway, S. 2009. *MATLAB: A Practical Introduction to Programming and Problem Solving.* New York: Elsevier.

Moler, C. B. 2004. *Numerical Computing with MATLAB.* Philadelphia: Society for Industrial and Applied Mathematics.

The MathWorks, Inc. 2007. http://www.mathworks.com.

Appendix C: Debugging a Simulink Model

Most of the debugging that we need to do for simple problems can be done just by catching block output values that are not correct. This is done most easily by using a Floating Scope block to display values at appropriate points in the model. But difficult problems or complicated, multilayer systems require a more powerful tool for locating bugs. This tool is available in Simulink® as the Debugger. In this Appendix we'll introduce setting up and running the Debugger on a simple model. For a more detailed discussion of the Debugger, read the Simulink documentation.

Let's use as our demonstration system the model shown in Equation (C.1).

$$x(t) = at + b \quad a = 10.5, b = 2.1 \tag{C.1}$$

This is a simple model that could easily be debugged with a Floating Scope, but we use it to demonstrate the Debugger operation.

C.1 STARTING THE DEBUGGER

We begin by starting Simulink and loading the demonstration model. When we have loaded the model, we see the model window in Figure C.1.

The first thing we must do is to open the Simulation | Configuration Parameters window and select the Optimization screen. Then clear the Block-reduction optimization and Signal-storage reuse checkboxes to turn off any optimization. If we look at the right-hand end of the toolbar above the window, we see the icon for the Debugger, as seen in Figure C.2 (find the cursor arrow). If we left-click this icon, the Debugger will open in a window of its own and should look like Figure C.3. The left-hand pane displays the list of breakpoints that have been set (none in Figure C.3), and the right-hand pane displays the Debugger output.

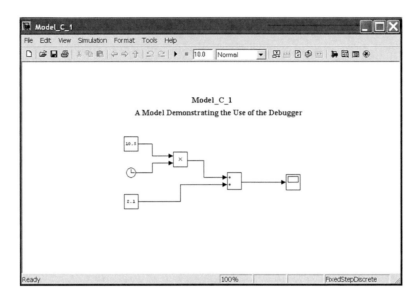

FIGURE C.1   Model_C_1 used for the Debugger demonstration.

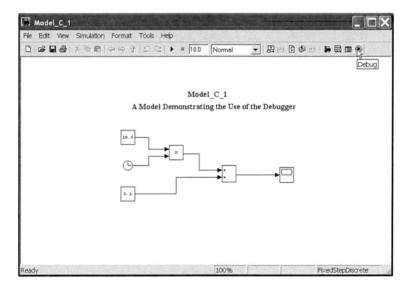

FIGURE C.2   Location of the Debugger icon on the Simulink toolbar.

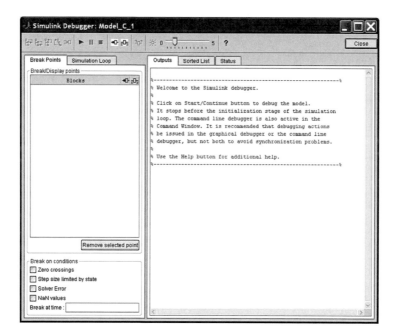

FIGURE C.3   The initial Debugger.

C.2 RUNNING THE MODEL IN STEP MODE

Let's run the simulation block by block and watch how it proceeds. We can do this by selecting the blocks whose execution initiation will cause the Debugger to pause and whose input and output values we would like to see at the pauses. We do this by selecting the block in the model window and then clicking the Block Breakpoint icon in the Debugger toolbar, as shown in Figure C.4 (find the cursor arrow).

Following the selection of the Block Breakpoint icon, an item appears in the Breaks/Displays list on the left-hand side with two boxes, of which the break-on-entry box has been selected. Returning to the model window, we can select a new block in this model and repeat the process. If we select all the blocks for breakpoint, we see the window shown in Figure C.5. We can see that list of breakpoints in the left-hand pane, and we note that the name of each breakpoint has the form Model/Name/BlockName.

Next, we ask the Debugger to give us input and output information when we execute the blocks by selecting each item in the Break/Display list and checking the boxes under the column headed by the Display Block I/O icon in Figure C.6.

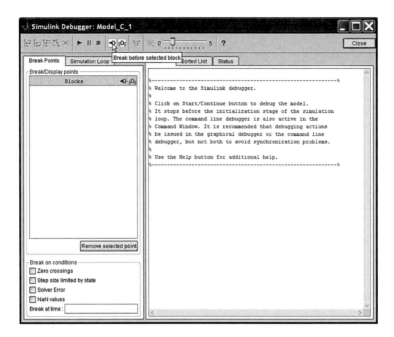

FIGURE C.4 Location of the block breakpoint icon on the Blocks Display toolbar.

FIGURE C.5 The Debugger window with blocks selected for Breakpoint.

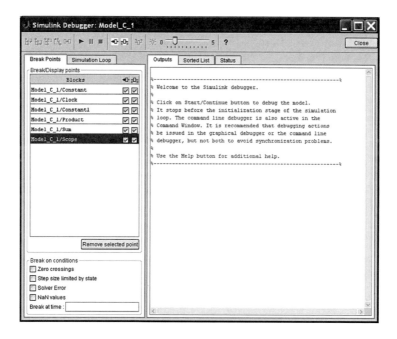

FIGURE C.6   The Debugger window with blocks selected for I/O display.

We double-click the Scope block so we can see the output as it develops, and we begin the execution by clicking on the Run icon in the toolbar of the Debugger window, as shown in Figure C.7.

This causes the Debugger to start running and pause at the first step. The Debugger window is then displayed at its initial position, as seen in Figure C.8.

The Simulation Loop pane is now at the front, showing the current state of the simulation loop. The right-hand column shows the block ID and position of the current breakpoint. In the Outputs panel, we see a trace packet begun by the border

$$(\% - \%)$$

and giving the current simulation time (TM = 0), the method name that the trace is describing (Mode_C_1.Simulate), and a Debugger internal pointer. Note in Figure C.9 that the Model window also reflects the first step by showing the debug title in the corner.

Clicking the Run button again causes the Debugger to execute the simulation up to the next breakpoint and pause with the output shown

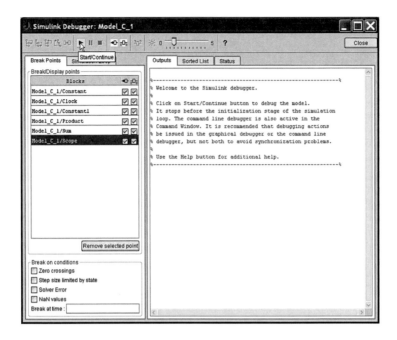

FIGURE C.7 Location of the Run icon on the Debugger toolbar.

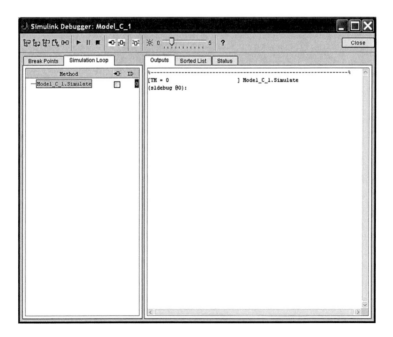

FIGURE C.8 The Debugger at entry.

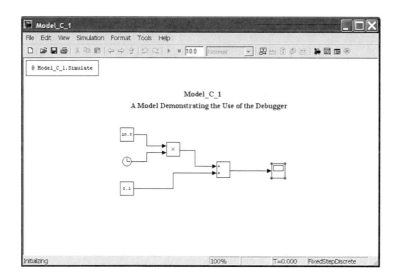

FIGURE C.9   The model at entry.

in Figure C.10. For us, that will be the first block to be executed. The Debugger has printed the output line

```
At break point:0 before 0:0 Constant block methods
'Model_C_1/Constant'
```

to show where it stopped, and the Display I/O has printed out its message. At this point, we are stopped just before executing the Constant block.

To see the order in which the blocks will be executed, we can select the Sorted List tab in the Output window, and it will show a list, as seen in Figure C.11.

The Model window shows points to the breakpoint location, as Figure C.12 shows.

If we continue to step through the simulation by repeatedly clicking the Run button, we will progress through the initializations and then arrive at the first set of executions for $t = 0.1$. Figure C.13 shows the state after enough clicks to compute the output at this point. If we look in Figure C.13 at the data for the Product block at $t = 0.1$, we can see that the Display I/O shows the lines

```
Trace: Data of 0:2 Product block 'ModelCp1/Product':
U1 = [10.5]
U2 = [0.10000000000000001]
Y1 = [1.05]
```

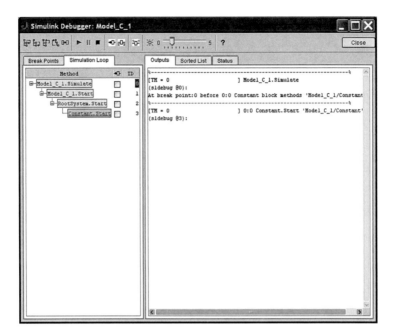

FIGURE C.10   (See color insert following page 144.) The Debugger at the first breakpoint.

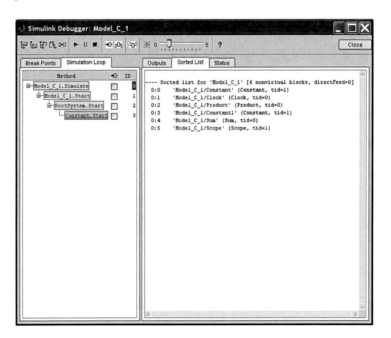

FIGURE C.11   The sorted breakpoint list.

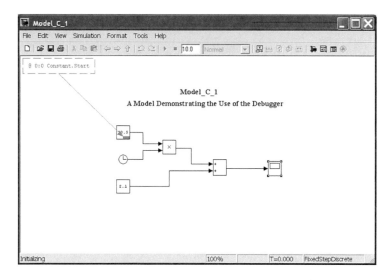

FIGURE C.12 (See color insert following page 144.) The model window showing the breakpoint location.

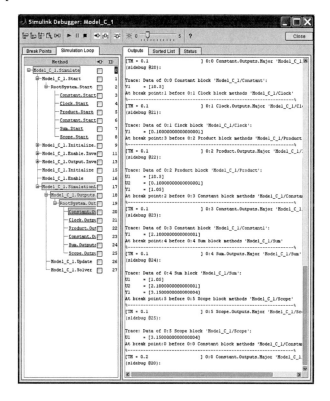

FIGURE C.13 The Debugger after $t = 0.1$.

The Un are the values of the inputs to the block, and the Yn are the values of the outputs of the block. This data enables us to verify correct operation of the block.

Note that the Scope output shows the first part of the curve close to the origin of the graph after it has been developed at the Scope block, as shown in Figure C.14. We can continue to step through the model simulation and observe the outputs of the blocks to verify that they are correct or that there is an error at some point.

While we have examined the Debugger only for a very simple model, our primary purpose has been to understand its use. Note that a large, complex simulation will likely require its use at some point. A good strategy to use for a large simulation is to develop a set of "known output" tests, so that correct operation of the simulation can be tested in detail using the Debugger.

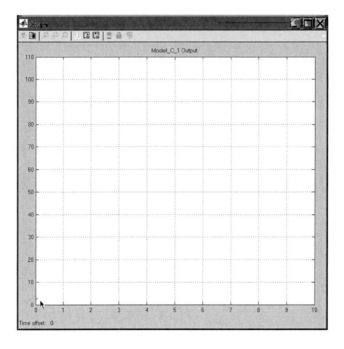

FIGURE C.14   The Scope output after $t = 0.1$.

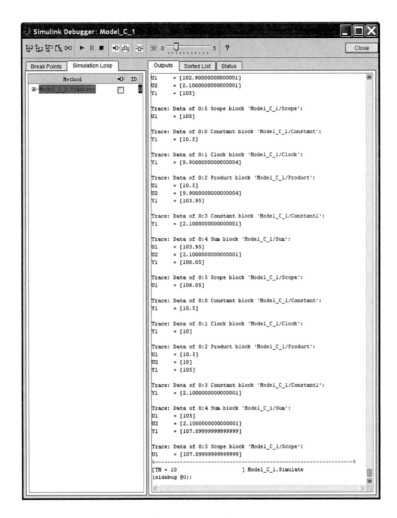

FIGURE C.15   The Debugger after a Trace Only run.

C.3 RUNNING THE MODEL IN TRACE MODE

We may want to run the simulation and gather the trace data without taking the time to step through the operation. We can do this in the windows shown in Figures C.15 and C.16 by selecting only the Display I/O checkbox (leaving the Breakpoint checkbox unchecked). Then there are no breakpoints at any blocks, and clicking the Run button causes the simulation to run to completion, leaving the I/O traces in the Outputs pane. Figures C.15 and C.16 show the Debugger and Model output after a Trace mode run with the example model.

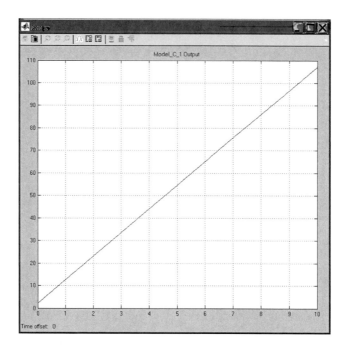

FIGURE C.16   The Model output after a Trace Only run.

ADDITIONAL READING

The MathWorks, Inc. 2007. http://www.mathworks.com.

Index